INTEGRATING SUSTAINABLE DEVELOPMENT INTO THE CURRICULUM

INNOVATIONS IN HIGHER EDUCATION TEACHING AND LEARNING

Series Editor: Patrick Blessinger

Previous Volumes:

Volume 1:	Edited by John M. Carfora and Patrick Blessinger
Volumes 2–4:	Edited by Patrick Blessinger and John M. Carfora
Volume 5:	University Partnerships for Community and School System Development – Edited by Barbara Cozza and Patrick Blessinger
Volume 6:	Emerging Directions in Doctoral Education – Edited by Patrick Blessinger and Denise Stockley
Volume 7:	University Partnerships for Academic Programs and Professional Development
Volume 8:	University Partnerships for International Development
Volume 9:	Engaging Dissonance
Volume 10:	University Partnerships for Pre-service and Teacher Development
Volume 11:	Refugee Education: Integration and Acceptance of Refugees in Mainstream Society – Edited by Enakshi Sengupta and Patrick Blessinger
Volume 12:	Contexts for Diversity and Gender Identities in Higher Education: International Perspectives on Equity and Inclusion – Edited by Jaimie Hoffman, Patrick Blessinger and Mandla Makhanya
Volume 13:	Strategies, Policies, and Directions for Refugee Education – Edited by Enakshi Sengupta and Patrick Blessinger
Volume 14:	Perspectives on Diverse Student Identities in Higher Education – Edited by Patrick Blessinger
Volume 15:	Language, Teaching and Pedagogy for Refugee Education – Edited by Enakshi Sengupta and Patrick Blessinger
Volume 16:	Strategies for Fostering Inclusive Classrooms in Higher Education – Edited by Jaimie Hoffman, Patrick Blessinger and Mandla Makhanya
Volume 17:	Strategies for Facilitating Inclusive Campuses in Higher Education: International Perspectives on Equity and Inclusion – Edited by Jaimie Hoffman, Patrick Blessinger and Mandla Makhanya

INNOVATIONS IN HIGHER EDUCATION TEACHING AND LEARNING VOLUME 18

INTEGRATING SUSTAINABLE DEVELOPMENT INTO THE CURRICULUM

EDITED BY

ENAKSHI SENGUPTA
Centre for Advanced Research in Higher Education, USA and International HETL Association, USA

PATRICK BLESSINGER
St. John's University, USA and International HETL Association, USA

TAISIR SUBHI YAMIN
International Association of Educators for World Peace, Germany

Created in partnership with the International Higher Education Teaching and Learning Association

Higher Education Teaching & Learning

https://www.hetl.org/

emerald PUBLISHING

United Kingdom – North America – Japan
India – Malaysia – China

Emerald Publishing Limited
Howard House, Wagon Lane, Bingley BD16 1WA, UK

First edition 2020

British Library Cataloguing in Publication Data
A catalogue record for this book is available from the British Library

ISBN: 978-1-78769-942-7 (Print)
ISBN: 978-1-78769-941-0 (Online)
ISBN: 978-1-78769-943-4 (Epub)

ISSN: 2055-3641 (Series)

Printed and bound by CPI Group (UK) Ltd, Croydon, CR0 4YY

ISOQAR REGISTERED
Certificate Number 1985
ISO 14001

ISOQAR certified Management System, awarded to Emerald for adherence to Environmental standard ISO 14001:2004.

INVESTOR IN PEOPLE

CONTENTS

PART II
INNOVATIVE APPROACHES

LIST OF CONTRIBUTORS

Luis Kluwe de Aguiar	Harper Adams University, UK
Meltem Alemdar	Georgia State University, USA
Anis Azizi	University of Auckland, New Zealand
Patrick Blessinger	International Higher Education Teaching and Learning Association, USA
Paul Benneworth	Western Norway University of Applied Sciences, Norway
Penny Brothers	Australian National University, Australia
Alice Cassidy	In View Educational Development, Canada
Ties Coomber	University of Auckland, New Zealand
Kata Dosa	University of Wisconsin–Madison, USA
Laura Franco Garcia	University of Twente, The Netherlands
Alana Griffith	University of the West Indies at Cave Hill, Barbados
Dawne Gurbutt	University of Central Lancashire UK
Russell Gurbutt	University of Bolton, UK
Euan Hague	DePaul University, USA
Niki Harré	University of Auckland, New Zealand
Jennifer Hirsch	Georgia Institute of Technology, USA
Robert Huddleston	Birmingham City University, UK
Ben Jongbloed	University of Twente, The Netherlands
Ellinor King	University of Auckland, New Zealand
Renze Kolster	University of Twente, The Netherlands
Martha Elena Núñez López	Tecnologico de Monterrey, Mexico
Roberto Pablo Martínez Lozano	Tecnologico de Monterrey, Mexico

Louise Manning	Royal Agricultural University, UK
Andrea Michelle Mead	University of Auckland, New Zealand
Winston Moore	The University of the West Indies at Cave Hill, Barbados
Michael W. Mulnix	American University of Iraq, Iraq
Sunni Newton	Georgia Institute of Technology, USA
Sarah Nyrose	Hawthorne Naturopathic Centre, Canada
Raghu Pucha	Georgia Institute of Technology, USA
Howard Rosing	DePaul University, USA
Sarah Saeckel	University of Stuttgart, Germany
Joseph P. Schwieterman	DePaul University, USA
Enakshi Sengupta	International Higher Education Teaching and Learning Association, New York, USA
Yona Sipos	University of Washington, USA
Martin Stienstra	University of Twente, The Netherlands
Manuel Vallée	University of Auckland, New Zealand
Taisir Subhi Yamin	International Centre for Innovation in Education, Germany
Samantha Zi Lin Yeo	University of Auckland, New Zealand
Ruthie Yow	Georgia Institute of Technology, USA
Yulun (Darren) Zhang	University of Auckland, New Zealand

SERIES EDITORS' INTRODUCTION

INNOVATIONS IN HIGHER EDUCATION TEACHING AND LEARNING

The purpose of this series is to publish current research and scholarship on innovative teaching and learning practices in higher education. The series is developed around the premise that teaching and learning is more effective when instructors and students are actively and meaningfully engaged in the teaching-learning process.

The main objectives of this series are to:

(1) present how innovative teaching and learning practices are being used in higher education institutions around the world across a wide variety of disciplines and countries,
(2) present the latest models, theories, concepts, paradigms, and frameworks that educators should consider when adopting, implementing, assessing, and evaluating innovative teaching and learning practices, and
(3) consider the implications of theory and practice on policy, strategy, and leadership.

This series will appeal to anyone in higher education who is involved in the teaching and learning process from any discipline, institutional type, or nationality. The volumes in this series will focus on a variety of authentic case studies and other empirical research that illustrates how educators from around the world are using innovative approaches to create more effective and meaningful learning environments.

Innovation teaching and learning is any approach, strategy, method, practice or means that has been shown to improve, enhance, or transform the teaching-learning environment. Innovation involves doing things differently or in a novel way in order to improve outcomes. In short, Innovation is positive change. With respect to teaching and learning, innovation is the implementation of new or improved educational practices that result in improved educational and learning outcomes. This innovation can be any positive change related to teaching, curriculum, assessment, technology, or other tools, programs, policies, or processes that leads to improved educational and learning outcomes. Innovation can occur in institutional development, program development, professional development, or learning development.

The volumes in this series will not only highlight the benefits and theoretical frameworks of such innovations through authentic case studies and other empirical research but also look at the challenges and contexts associated with implementing and assessing innovative teaching and learning practices. The volumes represent all disciplines from a wide range of national, cultural and organizational contexts. The volumes in this series will explore a wide variety of teaching and learning topics such as active learning, integrative learning, transformative learning, inquiry-based learning, problem-based learning, meaningful learning, blended learning, creative learning, experiential learning, lifelong and lifewide learning, global learning, learning assessment and analytics, student research, faculty and student learning communities, as well as other topics.

This series brings together distinguished scholars and educational practitioners from around the world to disseminate the latest knowledge on innovative teaching and learning scholarship and practices. The authors offer a range of disciplinary perspectives from different cultural contexts. This series provides a unique and valuable resource for instructors, administrators, and anyone interested in improving and transforming teaching and learning.

Enakshi Sengupta
Associate Editor,
International HETL Association

Patrick Blessinger
Founder, Executive Director, and Chief Research Scientist,
International HETL Association

FOREWORD

I have spent more than three decades in the field of education founding new universities and setting up their curriculum across the globe. I have found that in recent years the concept of sustainability has garnered interest among educators, mainly being associated with the field of business in relation to the practices of corporations. In most institutions of higher education, sustainability is not yet a mandatory requirement for undergraduate courses and is mainly imparted at the behest of individual faculty members who become the primary drivers of integrating corporate social responsibility and sustainability into undergraduate courses. As it is common in business practice, any sustainability initiative should be linked and defined with its outcomes and it is only then that the success factors can be measured. Deliverables of a course whether it is a module, or a workshop becomes the vehicles by which sustainability is evidenced and communicated to its target audience.

Why has it become necessary in recent years to impart sustainability education to the young learners and at times to the corporate world? If we are to analyze some simple choices in life, we realize that all human beings at any given point are consumers of products and services. Knowledge in sustainability helps individuals in their consumption and disposition choices. By imparting knowledge in sustainability, students are made aware of the subject and it potentially increases the students' future influence over the business community.

The world has seen unprecedented examples in the recent past of global warming in the form of cyclones and other atmospheric disasters. Global agencies concerned with the issues of energy-related climate change have emphasized ethics and sustainability education in curricula. Similarly, other global advising bodies have called for greater focus on sustainability in training and educating future business leaders. Sustainability practices emphasize the long-term "triple bottom line" (Elkington, 1998) as opposed to short-term shareholder view of profitability. While teaching sustainability issues to students, one needs to emphasize that the curriculum doesn't advocate charity giving or is opposed to profitability of an organization but an opportunity to revitalize companies and the economy through preserving and saving, rather than exploiting the environment and human social systems (Bradbury, 2003). The triple bottom line perspective helps an institution of higher education to integrate itself in the big picture view of the business value chain.

A "sustainable university" as noted by Ferrer-Balas et al. (2008) is beyond campus greening or tree plantation drives. Some of the notable features of a sustainable university includes transformative education, indulging and promoting transdisciplinary research, an approach toward solving societal issues and a university leadership and vision that promotes proactive responses to societal needs. In its societal problem-solving approach such universities need to inculcate in their curriculum active engagement with industry and other organizations where students can be involved in real problem solving. Such experiential learning should be included in the curriculum of sustainability which should involve active research along with engagement and partnership.

Sustainability paradigm requires to examine the values, beliefs, and assumptions of both educators and students and how we impact the environment and our economy. We try and inculcate in our young learners the respect for the planet and the resources that it gives us to survive. The process may not be simple and will be fraught with challenges. Yet, we all agree that the curriculum on sustainability can only be rendered through transformative learning practices using concrete case studies and global examples. This book will help expose the students to diverse viewpoints, understand the best practices in the world, and formulate viable solutions to complex problems.

The book talks about examples ranging from a postgraduate curriculum development to teach integrated care to clinical staff in UK to case studies exploring the comparative analysis of approaches to integrate sustainability into the curriculum in an university in the Caribbean. These case studies will help the readers understand the various approaches undertaken by educators to integrate sustainability into their curriculum. There is no one size fit all approach to it and designing a successful curriculum depends on ongoing learning and learning by doing which actively involves all stakeholders of the institution of higher education.

Dr. Michael William Mulnix
President of American University of Iraq,
Baghdad, Iraq

REFERENCES

Bradbury-Huang, H. (2003). Sustaining inner and outer worlds: A whole-systems approach to developing sustainable business practices in management. *Journal of Management Education, 27*(2), 172–187. https://doi.org/10.1177/1052562903251414

Elkington, J. (1998). *Cannibals with forks: The triple bottom line of 21st century business*. Stony Creek, CT: New Society Publishers.

Ferrer-Balas, D., Adachi, J., Banas, S., Davidson, C.I., Hoshikoshi, A., Mishra, A., ... Ostwald, M. (2008). An international comparative analysis of sustainability transformation across seven universities. *International Journal of Sustainability in Higher Education, 9*(3), 295–316. https://doi.org/10.1108/14676370810885907

PART I

IN PRACTICE

INTRODUCTION TO INTEGRATING SUSTAINABILITY INTO CURRICULUM

Enakshi Sengupta, Patrick Blessinger and Taisir Subhi Yamin

ABSTRACT

In this ever changing world, managing our ecosystem and creating a sustainable future seems to be one of the biggest challenges facing humanity. This challenge is further enhanced by ignorance or apathy of people toward the concept of sustainability. In most cases, students who are our future generation are left without any insight, commitment or even understanding their role and responsibility toward creating any meaningful beliefs and actions related to sustainability. Sustainability education is becoming crucial, mainly for young generation so that they have an understanding of concepts such as economic prosperity, resource equity, energy uses, and environmental health and concerns. While educating them on sustainability begins in institutions of education, it is important that sustainability education is well entrenched in the curriculum and everyday practice of their lives. This chapter introduces the volume series on sustainability where authors from different parts of the world narrate their own experience of imbibing sustainability into their curriculum and teaching sustainability to students.

Keywords: Sustainability; curriculum; future generation; environmental concern; education; economic prosperity; transformative education; interdisciplinary approach

Integrating Sustainable Development into the Curriculum
Innovations in Higher Education Teaching and Learning, Volume 18, 3–14

ISSN: 2055-3641/doi:10.1108/S2055-364120200000018018

INTRODUCTION

United Nations Educational, Scientific and Cultural Organization (UNESCO) (2009) advocates imparting knowledge in sustainability to students through which it enables students to develop the attitude, skills, and knowledge toward sustainability and take firm and decisive actions in their everyday lives to help inculcate sustainability throughout society. Education in sustainability has been supported by global frameworks such as the United Nations' Decade of Education for Sustainable Development (2005–2014) and the Global Action Programme on Education for Sustainable Development (post-2014), both led by the United Nations Educational, Scientific and Cultural Organization (UNESCO) (Buckler & Creech, 2014; UNESCO, 2014).

Higher education institutions have a role and responsibility in promoting sustainability in young minds by developing students as change agents (Moore, 2005; Svanstrom, Lozano-Garcia, & Rowe, 2008), encouraging ideas which are expressed freely, challenging paradigms, and promoting creativity and growth of knowledge (Barth, Godemann, Rieckmann, & Stoltenberg, 2007; Cortese, 2003; Lozano, 2006). Sustainability is not a standalone concept, and it is reflected in various other subjects, being multidisciplinary in nature (UNESCO, 2014). The very nature of this subject can be a contested terrain, which is multilayered and multifaceted, thus creating scopes for debates and creative learning (UNESCO, 2014).

It is argued and agreed by educationists that there exists an urgent need to educate students and enable them to understand sustainability matters so that while growing up they assume the role of pro-sustainability citizens, caring for their planet and surrounding with a futuristic view toward conserving and preserving the planet for future generations. In spite of having a general consensus about the positive effects of sustainable education there lies a whole plethora of contradiction surrounding the nature of pedagogy and the right way to administer sustainability education. It is often perceived in colleges and universities that sustainability-driven courses are delivered with a banking concept of education where a lecture-driven class often deposit knowledge without a proper understanding of the concept and without facilitating a process to address curiosity and encouraging debates. In light of such methods, it becomes imperative that new methodologies and frameworks are developed by academics to facilitate the understanding of sustainability (Cervantes, 2007). A worldview should be imparted to students that encompasses different interpretations of sustainability so as to avoid educational disconnect. Goekler (2003) argues that the concept of sustainability cannot be learned by merely receiving lecture-driven notes. Students need to develop the ability to think multifaceted ways and be tolerant to differing and even paradoxical views from different corners of the world.

The aim of imparting education to understand sustainability is to empower students to make decisions that balances the need to preserve healthy ecosystems and ultimately promote vibrant economies with an equitable social system. Studies have shown that students who are given discourses in sustainability are more motivated, better behaved, and sensitive toward their environment.

Such curriculum provides connections between students and their local communities facilitating collaborative work in promoting healthy lifestyles. Sustainability education has not become mandatory in many institutions of higher education and is often neglected as a component of undergraduate curricula in most institutions. Hence, individual faculty members from their own initiative and interest becomes the primary drivers of integrating sustainability education into undergraduate courses (Matten & Moon, 2004).

Accreditation and advising bodies in higher education have urged the pressing need to impart education focused on corporate social responsibility (CSR) and sustainability mainly to train future leaders (e.g., The Association to Advance Collegiate Schools of Business [AACSB] International, UNESCO, and the World Business Council for Sustainable Development). Yet, many institutions fail to realize the need for such training and still remain a firm believer of "shareholder value oriented governance of capitalist organizations" model (Matten & Moon, 2004, p. 329). Battle to reconcile such views on one hand and on the other hand promoting concepts of sustainability and that of social responsibility remains a challenging aim for academics.

Research into the subject of teaching sustainability investigated deans and directors at the top 50 global MBA programs (2006 Global MBA rankings, by *Financial Times*) revealed that a majority of them are required to teach ethics, CSR, and sustainability topics and have included them as a part of their curriculum (Christensen, Peirce, & Hartman, Hoffman, & Carrier, 2007). These topics and others subjects associated with them have seen a significant rise in incorporating them into mainstream curriculum in Western European countries and the United States. An institution can prepare its leaders for tomorrow to lead toward sustainability only after the students have been given a full exposure to the subjects with the help of their curriculum. "Undergraduate curriculum should be modified so every student in all academic departments has knowledge of the subject" (Kaifi, Khanfar, Noor, & Poluka, 2014, p. 40).

Higher educational institutions while creating future leaders for tomorrow "need to have a proactive strategy in shaping the perspectives of business leaders through sustainability related management education, research, management, and training programmes" (Park, Sarkar, & Bunch, 2012, p. 4).

Education has the capability of being a catalyst in changing and preparing students with capabilities to promote sustainability. In order to respond to these challenges, higher education institutions should undertake adequate steps in transforming not only their curriculum but also their teaching and learning techniques (Barth, 2014). Academics should develop programs and modules catering toward imparting lessons on sustainability and CSR and use innovative teaching methods that integrate topics like sustainability, sustainable development, CSR into their courses and curricula (Parkes et al., 2017).

Academics in the past have conducted studies to find out about courses on sustainability that are offered in universities. Their study revealed that a limited number of universities have imbibed these courses into their curriculum. Rundle-Thiele and Wymer (2010) had examined standalone courses in New Zealand and Australia and found that only 27% of universities in Australia offered a dedicated course in ethics, CSR, or sustainability. Exploration of both European and US-based institutions found that CSR and sustainability (CSRS) are offered as compulsory standalone courses or modules (Moon & Orlitzky, 2011). The data from these studies showed that more than 75% of undergraduate programs and more than 55% of MBA programs offered CSRS courses.

The importance of such courses is emphasized by accreditation bodies such as AACSB International through its Resource Centre for Business Ethics, as well as by the establishment of an ethics commission. While the center is focused on business ethics, the stated mission of the resource center is to impart a comprehensive source where information can be gathered on tools and frameworks and discussions can be made regarding ethics, sustainability, and CSR in business schools (AACSB International, 2013). AACSB International explicitly ties three concepts of ethics, CSR, and sustainability in one chain and advocates the three areas be integrated in all disciplines. Yet, no comprehensive study has been conducted to find out how individual topics have been presented and its impact on the learning outcome of the students.

LITERATURE REVIEW

Bridges and Wilhelm (2008) suggested three E's approach to teaching sustainability which includes ecology, equity (social), and economic (financial). Academics while teaching the concept was urged to consider that business success will depend on ecological and social success and a negative impact on either of them will show financial outcome in poor result.

AACSB, the accreditation board, has proposed a new standard while elaborating the teaching of sustainability and CSR and has explicitly stated that an institution of higher education must exhibits its commitment to emerging CSR issues. It should have an inclusive policy where the institution practices diversity, sustainable development, and environmental sustainability across cultures. These concepts should be further embedded in its procedures, curricula, research, and other activities.

Education for sustainability is understood differently by different faculty members. This ever evolving subjects with various dimensions and definitions requires academics who are keen on transformative education urging a personal transformation if required (Wooltorton, 2002). While working on students' comprehension of their learning about sustainability, Segalàs, Ferrer-Balas, and Mulder (2010) found that student's comprehension of the term sustainability is confined to technology and discovered minimum relevance in social and attitudinal aspects.

A second finding in this field showed that a student only exhibited knowledge and understanding about sustainable development which applied a more community-oriented and constructive approach (Myers & Beringer, 2010). In a more recent study, Segalàs, Mulder, and Ferrer-Balas (2012) discovered a "mismatch" that lies among the "experts" and students' understanding of sustainability. Students either perceive sustainability as unrelated to social and institutional aspects or they barely perceive sustainability as a complex issue; hence, a general apathy can be seen in studying or trying to understand the subject. Researchers in this field have urged on systems and multidisciplinary thinking, with a greater focus toward, "reorientation of the pedagogy and the learning processes" (Segalàs et al., 2012, p. 302).

Sustainability teaching is not confined to books and lecture slides but requires both educators and students to understand their own values, hidden assumptions, motivations, beliefs, and actions (Holdsworth, Wyborn, Bekessy, & Thomas, 2008). Teachers cannot merely teach the subject without being involved in it and reflecting how their own work, their knowledge, and their teaching may impact the environment and economy of the region they belong to, especially if it is a growing and transforming economy. Active and reflective learning with transformative and participative approach with the usage of case studies developed locally and globally help to expose the students to various viewpoints and urge them to think critically in solving complex global issues and consider the consequence of their acts and accept responsibility for creating a sustainable future for their next generation (Scott, 2009).

While presenting his case on sustainability in higher education, Sterling (2004) puts forward three approaches that an educational institution can adopt while teaching sustainability:

(1) Educating about sustainability – an accommodative response.
(2) Education for sustainability – a reformative response.
(3) Capacity building – a transformative response.

The first and the most basic level is to offer modules on sustainability with the rest of the curriculum designed for the school. The next level is more advanced which talks about the institution transforming itself as a sustainable organization. The third level is more comprehensive in dealing with students who can be transformed by adopting the skills of sustainability. Starik, Marcus, and Clark (2010, p. 377) criticize the "incrementalism reform approaches that most individuals, organisations and societies have employed to address critical global sustainability issues…" and demand more transformative sustainability results in higher education, mainly while teaching management education.

Stubbs and Schapper (2011) advocated interdisciplinary course content while including sustainability education into the curriculum. Beijing Normal University in China and Aalborg University in Denmark added another dimension toward sustainability education by adopting interdisciplinary and cross-cultural content in a project-based learning approach toward sustainability (Du, Su, & Liu, 2013). Chhokar (2010) pointed it out that most institutions of higher education lacks interdisciplinary skills of staff and students which is needed to integrating sustainability into the curriculum.

CONCLUSION

Promoting sustainability with the help of curriculum is by no means an easy task as one needs to incorporate the take-away factor of the student while designing the course. System and consistent framework to assess the learning experience and learning outcome has to be kept in consideration at all times while designing such courses. The syllabus, learning modules, assessment protocol are vehicles through which sustainability is communicated and imbibed into the learners. While designing a curriculum with modules on sustainability the faculty members have to keep in mind the trends that impact a learner's success. The instructor's knowledge and passion about the subject shall determine the instructor–learner relationship and the outcome of the delivery of the modules. This relationship often transcends the content-specific competencies and results in sustainability in action where the instructors have observed the learners keenness to apply sustainability in everyday life.

CHAPTER OVERVIEWS

"Integrating Sustainable Development into Healthcare Curriculum," by Professor Russell Gurbutt and Professor Dawne Gurbutt, sets the scene for the need of sustainable development in the healthcare curriculum by discussing the contemporary context of healthcare provision and its associated challenges. Set within the context of the developing contemporary healthcare agenda in the UK, the chapter focuses on an exploration of the drivers for integrated care and the impact this has on curriculum development. An example of post-graduate curriculum development to teach integrated care to clinical staff guides the reader through the issues presented by the context and the need for sustainability in both content and design. The challenges of barriers and enablers to integrated care is considered and the extent to which a cultural mind-set transition might occur among participants to align initiatives to sustainable development goals (SDGs). The curriculum required participants to critically understand the political, economic, social technological, and ethical contexts driving service redesign and apply critical thinking to navigate development paths through the changing health and social care landscape. A leadership dimension championed innovation applied to reimagining integrated responses to wicked problems.

Promoting sustainability veered away from accepting short-termism in favor of interventions that equip participants with the agility to address wicked problems and anticipate the rapid role change and development which is inherent within the service. The role of collaborative education in relation to embedding cross-cutting themes and patient voices as well as securing multiple gains in interdisciplinary themes, building networks, role modeling, and creativity thereby supporting a way of working that supports sustainability has been discussed in this chapter.

"Embedding Sustainable Development in the Curricula: Learning About Sustainable Development as a Means to Develop Self-awareness," by Louise Manning and Luis Kluwe de Aguiar, addresses the extent to which sustainable development can be taught as part of the curricula at a university in the United Kingdom (UK). Since the topic tends to be shrouded in complexity (Al Rawahy, 2013, p. 400), sustainable education is generally a difficult concept to explore in an educational setting. Those attempting to teach it ought to have enough inter- and transdisciplinary skills to be able to foster in the students not only knowledge assimilation but also deep acquisition of values (Lambrechts, Mulà, Ceulemans, Molderez, & Gaeremynck, 2013; Faham, Rezvanfar, Mohammadi, & Nohooji, 2017). Sustainable education is embedded in the curriculum of Food Technology students throughout their four year degree at Harper Adams University. A brief description of how curricula is organized is presented. Emphasis is given to one module in the final year of the degree where the content and pedagogical practice as well as the pedagogic nature of the learning environment is analyzed in the light of the literature.

Particularly, the learning outcomes suggested for sustainability-related subjects (Fields, 2009; OECD, 2008) and the works of Lambrechts et al. (2013) among others are used to scrutinize both the extent the teaching and learning practices are developed, and the extent to which they equip the students with skills such as critical thinking, analysis, reflection, and complex problem solving. Two case studies are presented: the Loess Plateau (China) and the Chiapas Coffee Farmers (Mexico). These serve to also contextualize the extent to which the students can draw from teaching and learning strategies in the module in question as well as other experiences (personal or from other modules), and use these to frame the problem-based learning. This allows them to gain a deeper meaning of the topics covered and also to work with problems with aspects of uncertainty and imperfect levels of information.

"A Comparative Analysis of Approaches to Integrating Sustainability into the Curriculum at a University in a Small Island Developing State in the Caribbean," by Alana Griffith and Winston Moore, is about the University of the West Indies (UWI) which has undoubtedly played a significant role in the development of the Caribbean. As a higher education institution in the Caribbean, the 70-year old UWI has accomplished this directly and indirectly by "advanc[ing] learning, creat[ing] knowledge and foster[ing] innovation for the positive transformation of the Caribbean and wider world." This chapter utilizes courses in different disciplines – economics and sociology – to highlight innovative teaching strategies that are employed to engage students, the inclusion of sustainability into the curriculum and linking taught units to the SDGs as well as how students confront these issues. The case study courses are delivered at the UWI's Cave Hill Campus in Barbados and the two courses are "Caribbean Social Problems" and "Economic Planning." The chapter demonstrates the cross-cutting nature of challenges to sustainable development needed in Caribbean Small States that are vulnerable to climate change, natural disasters, and other environmental challenges as well as the necessity of integrating sustainability and the SDGs into course delivery to ensure the future development of the region does not compromise future generations.

"Integrating Sustainable Development into the Curriculum: A Case Study on the Developing of Sustainability Competencies in Industrial Design Students at a Bachelor Level in Mexico," by Martha Elena Núñez López, Robert Huddleston, and Roberto Pablo Martínez Lozano, presents a case study on integrating sustainable development into a bachelor's degree course in industrial design.

The research is being conducted at Tecnologico de Monterrey (TEC), Mexico, where the lead author is a Professor of architecture. TEC is one of the world's largest universities, with 26 campuses nationwide and 91,285 students. Mexico has a five-year national development plan: the "Plan Nacional de Desarrollo 2013–2018" (Government of the Mexican Republic, 2013). It provides a basis for guiding the policies and programs of the Government of Mexico. The focus of this study is the PND's "quality education goal to make scientific, technological and innovation development pillars for sustainable economic and social progress."

The case study at TEC investigates a curriculum intervention, utilizing interviews with students to gather data on their responses to the university's current development of sustainability competencies. Their responses are explored through comparing a traditional semester with a semester in which sustainability contents and assessment criteria were added to the curriculum of the industrial design workshop courses. The results reveal that the students recognized a significant advance in their recognition and development of sustainability competencies and that this had resulted from this curriculum intervention. The chapter concludes by proposing that the findings of the study indicate that this pedagogical approach has the potential to contribute significantly to sustainable development education in Mexico.

Alice Cassidy, Yona Sipos, and Sarah Nyrose in their chapter "Programs, Workshops, Resources, and Other Supports for Post-secondary Sustainability Educators," speaks about the growing need to train and support educators to introduce or enhance aspects of sustainability into the curriculum. The authors have provided an overview of selected literature on the integration of sustainability development and education into the curriculum and related institutional leadership at the post-secondary level. Turning to educational development for sustainability education, they have shared tools and resources to support educators from any discipline, to introduce, integrate and/or enhance sustainability in their course, program or initiative. The authors found very few examples of workshops to help faculty and others who teach at colleges and universities. For one such example, the Sustainability Education Intensive (SEI), a three-day workshop was designed and led at the University of British Columbia (UBC). The chapter authors summarized the workshop aspects that they found helpful, and how participation affected them as sustainability educators. The authors further encourages post-secondary institutions to provide support in the form of workshops, tools, resources, and funding to help educators introduce or enhance aspects of sustainability into their courses and programs. Students are asking for this, and, as they are future leaders, it is important that educators address the numerous environmental, social, and economic issues that demand attention.

"*Hortus in Urbe*: Building a Sustainable Development Curriculum in Chicago," by Euan Hague, Howard Rosing, and Joseph P. Schwieterman, describes the development of an interdisciplinary graduate program focusing on sustainable urban development at Chicago's DePaul University. Locating the curriculum both in the administrative institutional context and the historical geography of a racially and economically segregated urban area, the authors discuss the process of program formation and adjustment over its first five years of operation, 2013–2018. The chapter highlights some of the challenges encountered by program faculty, from internal curricular competition to external classification of the program under federal educational designation and also highlights some of the interdisciplinary innovations, such as requiring courses in geographic information systems (GIS) to aid spatial data analysis and visualizations. In the second part of the chapter, to assess the impact of the graduate program on students, the authors review and draw from reflection essays written by students who, while completing their studies, pursued internships with nonprofit organizations in Chicago.

They suggest that the interdisciplinary learning delivered in the curriculum has led to local stakeholders rethinking and conceptualizing how they advance their sustainability goals. The chapter concludes by tabulating the current employment of recent alumni to indicate how the graduate program is equipping students with the skills to succeed and impact on sustainability policy and practice. The authors maintain that it is necessary to push the understanding of "sustainability" beyond solely environmental concerns. By widening the concept to incorporate understandings of how economic development and community engagement must be included to deliver a sustainable city, the interdisciplinary curriculum challenges students to become leaders in local effort to make urban areas not only greener, but also more economically and socially sustainable for all residents.

"An Interdisciplinary Teaching Module on the Global Clothing Industry: Lessons from Working Across Four Disciplines and Two Universities," by Niki Harré, Anis Azizi, Penny Brothers, Ties Coomber, Ellinor King, Andrea Michelle Mead, Sarah Saeckel, Manuel Vallée, Samantha Zi Lin Yeo, and Yulun (Darren) Zhang, presents a case study of The University of Auckland. As learning institutions that offer high-level expertise across a range of disciplines, universities are well positioned to develop students' understanding of the complex, multi-level issues contemporary societies face and how to address these. Since 2016, the university has been teaching an interdisciplinary module on the global clothing industry to students enrolled in an introductory psychology course and a second year chemistry course at The University of Auckland in New Zealand. In 2016, the module also involved third year chemistry students, and in 2017 second year sociology students and graduate students in English literature from the University of Stuttgart in Germany took part.

The module has the following features: (1) it focuses on a complex industry with ramifications for social and environmental sustainability; (2) it involves an issue of direct relevance to the students; (3) students teach those from another discipline as "subject experts," and (4) students are assessed on their learning within their home course. An evaluation of the 2018 iteration with psychology and chemistry students ($N = 185$) showed post-test decreases in participants' materialistic values and increases in knowledge and concern about the social and environmental impacts of the clothing industry. Many evaluation participants indicated that the module changed their decision-making when buying clothes, and there was a mean decrease in the frequency with which they intended to make purchases. The authors discussed the institutional barriers they faced and provide five recommendations for other university teachers considering integrating an interdisciplinary sustainability module into existing courses.

"Integrating Sustainability into a Freshman-Engineering Course Through an Institute-level Initiative: A Teaching–Learning Model with Authentic Activity and Context," by Raghu Pucha, Kata Dosa, Sunni Newton, Meltem Alemdar, Ruthie Yow, and Jennifer Hirsch, speaks about a campus-wide academic initiative ("Center for Serve-Learn-Sustain") which was launched in January 2016 at Georgia Tech. It was aimed at preparing undergraduate students in all majors to use their disciplinary knowledge and skills to contribute to the major societal challenge of creating sustainable communities. The initiative collaborates with faculty in all six Georgia Tech colleges to develop courses and cocurricular opportunities that will help students learn about sustainability and community engagement and hone their skills by engaging in real-world projects with nonprofit, community, government, and business partners. Affiliated courses address various aspects of the Centre's sustainable community's framework, which presents sustainability as an integrated system connecting environment, economy, and society, and focuses especially on society – particularly social equity and community voice – to address this gap in technological education.

This chapter reports on one engineering instructor's ongoing activities that bring social justice and sustainability into the engineering classroom through sociotechnical project-based learning. This cornerstone design course is one of more than 100 Centre-affiliated courses currently being offered, enrolling over 5,000 students across all six colleges. The activities in the course pertain particularly to the initiative's vision that all graduates of the institute – a majority of whom will graduate with engineering degrees – are able to contribute to the creation of sustainable communities and to understand the impact of their professional practice on the communities in which they work. A situated knowledge and learning pedagogical theory is used in the affiliated course, where concept, authentic activity, and context are involved in student learning to produce useable robust knowledge. Sociotechnical project-based teaching model with contextualized design problems are used to engage students throughout the course that incorporate social justice and environmental sustainability through both individual and team projects. Pedagogical approaches to learning, strategies, and challenges for the implementation and assessment of intervention activities, data analysis on student reflection, pre- and post-survey data analysis are presented.

"Integrating Sustainable Development into the Curriculum: Enacting "Scalar Shifting" in ESD Competencies," by Paul Benneworth, Renze Kolster, Martin Stienstra, Laura Franco Garcia, and Ben Jongbloed, talks about increasing interest in integrating sustainable development into higher education curricula in systematic ways to increase the sense of autonomy and agency of young graduates in addressing "grand challenges." This education for sustainable development (ESD) is not merely about providing knowledge of the challenges in delivering sustainable societies but in raising awareness of opportunities to create local solutions, a willingness or desire to construct those solutions, and the organizational skills to implement these solutions in context. However, because these courses are integrated into academic curricula, students must learn practical-ends driven skills in ways compatible with existing academic standards frameworks oriented toward theoretical understanding. These can lead to very different pedagogical orientations (theoretical and practical), one reason that could explain the relatively limited uptake of ESD within higher education to date. We develop a model by which a single educational experience could help to bridge between these two orientations and highlight that the key issue here is the transitions between the different scales of the problem, and how that in turn relates to the development of key ESD competencies. We explore this conceptual framework with a single study of an example of student volunteer projects in which students spend 2–4 months working on a knowledge transfer project to the global south, while also researching how that project contributes to solving the SDGs. We reflect on tensions, problems, and solutions to ensure that universities produce graduates who are effectively oriented to tackling the urgent contemporary societal issues, while also gaining valuable personal development experience and career perspectives.

REFERENCES

AACSB International. (2013). Ethics/sustainability resource center. Retrieved from http://aacsb.edu/resources/ethics-sustainability.about.asp

Barth, M. (2014). *Implementing sustainability in higher education: Learning in an age of transformation*. Oxford: Routledge.

Barth, M., Godemann, J., Rieckmann, M., & Stoltenberg, U. (2007). Developing key competencies for sustainable development in higher education. *International Journal of Sustainability in Higher Education*, *8*(4), 416–430.

Bridges, C. M., & Wilhelm, W. B. (2008). Going beyond green: The "why and how" of integrating sustainability into the marketing curriculum. *Journal of Marketing Education*, *30*, 33–46.

Buckler, C., & Creech, H. (2014). *Shaping the future we want. United Nations Decade of Education for Sustainable Development (2005–2014). Final report. Decade of education for sustainable development monitoring and evaluation*. Paris, France: UNESCO.

Cervantes, G. (2007). A methodology for teaching industrial ecology. *International Journal of Sustainability in Higher Education*, *8*(2), 131–141.

Chhokar, K. B. (2010). Higher education and curriculum innovation for sustainable development in India. *International Journal of Sustainability in Higher Education*, *11*(2), 141–152.

Christensen, L. J., Peirce, E., Hartman, L. P., Hoffman, W. M., & Carrier, J. (2007). Ethics, CSR, and sustainability education in the Financial Times top 50 global business schools: Baseline data and future research directions. *Journal of Business Ethics*, *73*, 347–368.

Cortese, A. (2003). The critical role of higher education in creating a sustainable future. *Planning for Higher Education*, *31*(3), 15–22.

Du, X. Y., Su, L., & Liu, J. (2013). Developing sustainability curriculum by using the PBL method in a Chinese context. *Journal of Cleaner Production*, *61*, 80–88.

Faham, E., Rezvanfar, A., Mohammadi, S. H. M., & Nohooji, M. R. (2017). Using system dynamics to develop education for sustainable development in higher education with the emphasis on the sustainability competencies of students. *Technological Forecasting and Social Change*, *123*, 307–326.

Field, A. (2009). *Discovering statistics using SPSS*(3rd ed.). London: Sage Publications.

Goekler, J. (2003). Teaching for the future: Systems thinking and sustainability. *Green Teacher*, *70*(Spring), 8–14.

Government of the Mexican Republic. (2013). *Plan Nacional de Desarrollo*, 2013–2018, Mexico (pp. 17, 19, 22).

Holdsworth, S., Wyborn, C., Bekessy, S., & Thomas, I. (2008). Professional development for education for sustainability: How advanced are Australian universities? *International Journal of Sustainability in Higher Education*, 9, 131–146.

Kaifi, B. A., Khanfar, N. M., Noor, A. O., & Poluka, L. (2014). International business students' understanding, perception, and commitment to corporate social responsibility: A study based upon gender, generational affiliation, and culture. *Business and Management Research*, *3*(3), 34–42.

Lambrechts, W., Mulà, I., Ceulemans, K., Molderez, I., & Gaeremynck, V. (2013). The integration of competences for sustainable development in higher education: An analysis of bachelor programs in management. *Journal of Cleaner Production*, *48*, 65–73.

Lozano, R. (2006). Incorporation and institutionalization of SD into universities: Breaking through barriers to change. *Journal of Cleaner Production*, *14*, 787–796.

Matten, D., & Moon, J. (2004). Corporate social responsibility education in Europe. *Journal of Business Ethics*, *54*, 323–337.

Moon, J., & Orlitzky, M. (2011). Corporate social responsibility and sustainability education: A trans-Atlantic comparison. *Journal of Management & Organization*, *17*, 583–603.

Moore, J. (2005). Seven recommendations for creating sustainability education at the university level: A guide for change agents. *International Journal of Sustainability in Higher Education*, *6*(4), 326–339.

Myers, E. M., & Beringer, A. (2010). Sustainability in higher education: Psychological research for effective pedagogy. *Canadian Journal of Higher Education*, *40*(2), 51–77.

OECD. (2008). Workshop on education for sustainable development, 11-12 September 2008. Paris, France. Retrieved from http://www.oecd.org/greengrowth/41078703.pdf

Park, J., Sarkar, R., & Bunch, R. (2012). Sustainability and management education in China and India: Enabling a global green economic transition. *Business Leadership Review*, *9*(1), 1–15.

Parkes, C., Buonon, A. F. & Howaidy, G. (2017). The Principles for Responsible Management Education (PRME): The first decade – What has been achieved? The next decade – Responsible Management Education's challenge for the Sustainable Development Goals (SDGs). *The International Journal of Management Education*, 15, 61–65.

Rundle-Thiele, S. R., & Wymer, W. (2010). Stand-alone ethics, social responsibility, and sustainability course requirements. *Journal of Marketing Education*, *32*, 5–12.

Scott, R. H. (2009). Sustainable curriculum, sustainable university. *eCULTURE*, *2*, Article 15.

Segalàs, J., Ferrer-Balas, D., & Mulder, K. (2010). What do engineering students learn in sustainability courses? The effect of the pedagogical approach. *Journal of Cleaner Production*, *18*(3), 275–284.

Segalàs, J., Mulder, K., & Ferrer-Balas, D. (2012). What do EESD "experts" think sustainability is? Which pedagogy is suitable to learn it? *International Journal of Sustainability in Higher Education*, *13*(3), 293–304.

Starik, R., Marcus, & Clark. (2010). Editorial: In search of sustainability in management education. *Academy of Management Learning & Education*, *9*(3), 377–383.

Sterling, S. (2004). Higher education, sustainability and the role of systemic learning. In P. Corcoran & A. Wals (Eds.), *Higher education and the challenge of sustainability: Contestation, critique, practice, and promise*. Dordrecht, The Netherlands: Kluwer Academic.

Stubbs, W., & Schapper, J. (2011). Two approaches to curriculum development for educating for sustainability and CSR. *International Journal of Sustainability in Higher Education*, *12*(3), 259–268.

Svanstrom, M., Lozano-Garcia, F. J., & Rowe, D. (2008). Learning outcomes for sustainable development in higher education. *International Journal of Sustainability in Higher Education*, *9*(3), 339–351.

United Nations Educational, Scientific and Cultural Organization (UNESCO). (2009). *UNESCO world conference on education for sustainable development*. Paris, France: UNESCO.

United Nations Educational, Scientific and Cultural Organization (UNESCO). (2014). *UNESCO roadmap for implementing the global action programme on education for sustainable development*. Paris, France: UNESCO.

Wooltorton, S. (2002). *Education for sustainability: A background paper prepared for the state sustainability strategy*. Perth, Australia: Department of the Premier and Cabinet.

CHAPTER 1

INTEGRATING SUSTAINABLE DEVELOPMENT INTO HEALTHCARE CURRICULUM

Russell Gurbutt and Dawne Gurbutt

ABSTRACT

This chapter will set the scene for the need of sustainable development in the healthcare curriculum by discussing the contemporary context of healthcare provision and its associated challenges. This is specifically in the context of wicked or complex problems. An exploration of what gives rise to such problems is used to lead into integrated care as a proposed solution and its associated drivers. This is considered in relation to sustainability and curriculum development with a focus on two examples: the creation of a post-graduate curriculum to teach sustainable integrated care to clinical staff and a new pedagogical approach called socially immersive learning. Through this the challenges of barriers and enablers to integrated care will be considered and the extent to which a global mind-set transition might occur among participants to align sustainable development goals.

Keywords: Curriculum design; sustainability; interdisciplinary working; integrated care; healthcare; cultural mindset

INTRODUCTION

In 2018, the UK celebrated 70 years of the National Health Service (NHS), a service funded through direct taxation which was designed to be universal and free at the point of use. Its genesis was in the planning undertaken during the Second World War for a better post-war provision for the British population

Integrating Sustainable Development into the Curriculum
Innovations in Higher Education Teaching and Learning, Volume 18, 15–24

ISSN: 2055-3641/doi:10.1108/S2055-364120200000018019

(The National Archives, 1942). The intervening years have brought changes which Aneurin Bevan, the principal policy leader of the development, could not have foreseen. In 1948, there was a belief that the service could eventually be predicated upon "health" and its promotion rather than "illness" and its resolution. In 1948, the NHS provided basic medicine, surgery, and preventative care. Modern healthcare includes a broader range of services and interventions that have been enabled by advances in science and technology which could not have been anticipated 70 years ago. Not only has this raised expectations of what the service can and should provide, but it has also witnessed a steady increase in demand and a changing demographic that has seen the life expectancy of an adult male in England change from 67 years in 1948 to 83 years in 2010 (Office for National Statistics (ONS), 2018a). A consequence of a changing population profile in which the retired section is living longer and experiencing the co-morbidities of older age, a smaller working tax-paying part of the population and enhanced survival rates across a wide range of conditions and illnesses has led to increasing healthcare costs. A growing, diverse population adds to the complexity of healthcare provision. The UK is not alone in facing these issues; similar challenges will be faced across the developed world, and more complex ones in developing countries (World Health Organization (WHO), 2015).

These challenges are played out in political discourses around the funding of healthcare, a cultural backdrop of adherence to the principle of universality, and the increasing scope of technology. Narratives include different approaches to the consideration of healthcare from Gawande (2014) and to the consideration of alternative models with which to fund it (World Health Organization (WHO), 2017). These are not the only problems with healthcare provision, again, drawing on examples from the UK, it could be argued that there have been significant shifts in the way in which a career in healthcare is perceived, making it a less attractive option. One way in which a shift in social perception has been identified was illustrated in 2005 when the Royal College of Physicians described Medicine as a "vocation" which resonated strongly with many in the profession. However, necessary measures designed to protect the working experience of practitioners, such as the 2012 European Working Hours' Time Directive, have drawn attention to an increasing social emphasis on work–life balance; although medicine and healthcare remain exceptions to the concept of a regularized, working week (Neivas & Thaver, 2015). The social shift away from the expectations of a vocation to a work role coupled with a broader range of career options open to young people on leaving full-time education has posed a challenge for recruitment and retention into healthcare careers. This has been compounded through less competitive pay and conditions, and specifically public sector austerity measures that have kept wages largely static for over a decade (Office for National Statistics (ONS), 2018b). A further step taken to seek to increase the number of applicants into nurse education in 2017 has been to remove the bursary which paid tuition fees for eligible students. This translates into newly registered staff starting their careers with a debt in the context of less competitive wages to other employment sectors. Moreover, it has added to the difficulties for recruiters to programs, given that nursing students are on placement for 50% of their program as well as 50% taught contact hours, making it difficult to maintain paid work alongside study. A change in the expectations of work in health professions has also shaped the extent to which employers need to accommodate different patterns of working. The rise in the number of women studying medicine, while an important step in achievement of equality has also been accompanied by an increase in demand for part time working as parents juggle family commitments and career (Moberley, 2018).

It can be argued that the healthcare workforce has experienced a significant range of pressures due to policy shifts, social and technological change, and economic pressures that have all converged to create a complex problem that has seen enduring workforce shortages amidst rising demands. It could therefore be argued that the already complex UK health and social care service has undergone incremental, evolutionary change and has now reached a significant change point. The impact of the economic crisis of 2008 has significantly exacerbated pressures on providing socially responsible services.

The drivers of change have included real term funding cuts, increasing expectations and demands, managerialism, marketization, policy, litigation, and the continually evolving and costly technology developments offering a widening range of investigations and interventions.

The interconnected and interrelated challenges of a contemporary healthcare system can be described as a "wicked problem"; a term used to describe a problem that is difficult or impossible to solve, usually due to complexity and contradictory elements that are difficult to deconstruct or recognize (Hornbeek & Peters, 2017). The so-called wicked problem relates to the difficulties in resolving the problem rather than the inherent nature of the problem content. In the case of health services, there is a complex equation of increasing service quality, reduced costs, pursuing safe outcomes, policy interventions, recentering services around patents and managing people resources, and providing ongoing education to keep abreast of changes. However, within the context of austerity, service fragmentation and the accompanying loss of a national response, policy makers have increasingly seen integrated care as a solution, thus signaling the greatest extensive reform since the NHS' inception in 1948.

This move toward integrated care involves organizations integrating vertically (i.e., different organizations merging and aligning operations to new agreed goals) and horizontally (referring to the operational work processes that make for a smoother and efficient patient journey). In England, this has been led through by a Department of Health Policy (NHS, 2015) in the context of 10-year development plans (NHS England: Long term plan) and has progressed through a group of early adopter organizations classified as Vanguards (NHS England, 2015). Allied to this development in 2016 has been the creation of 44 regional sustainability and transformation partnerships covering the whole of England with an aim of establishing partnerships which could:

> evolve to form an integrated care system, a new type of even closer collaboration. In an integrated care system, NHS organisations, in partnership with local councils and others, take collective responsibility for managing resources, delivering NHS standards, and improving the health of the population they serve. (NHS England, 2016)

The intention of creating better and more "joined up" services for patients, with greater collaborative working among a range of local providers (including the third sector) based on public health data about specific contexts, has to be interpreted in terms of what had preceded this change. The period from the inception of the NHS until 2015 has seen successive reforms and measures intended to manage expenditure, drive up quality, increase internal competition, introduce private sector tendering for service provision, and improve efficiency via redesigning the workforce and operational processes. Politically led metrics purporting to demonstrate healthcare organization efficiency and effectiveness in terms of outcomes has led to a range of "league tables" that allow comparisons between provider organizations. This has led to criticism that it moves the focus away from care (Frances, 2013).

Such national changes, which arguably might be interpreted as the most far reaching service reform since 1948, mirror responses to similar wicked problems in other nations. Given that English workforce deficits are being addressed through steps to attract professionals from other countries, a global mind-set has to be taken to consider what promotes sustainable development as nations strive to provide services and support to promote health and wellbeing.

In 2015, the United Nations (UN) established 17 sustainable development goals (SDGs) that provide a useful framework against which to consider progress. Sustainable development goal 3 (SDG3) is to "ensure healthy lives and promote well-being for all ages." For this to be achieved, there needs to be: "universal health coverage, including financial risk protection, access to quality essential health-care services and access to safe, effective, quality and affordable essential medicines and vaccines for all" (SDG3). Integrated care relates to SDG17 which is to "strengthen the means of implementation and revitalise the global partnerships for sustainable development." However, none of this can be truly effective without it hinging on partnerships and collaborations through "a shared vision, and shared goals that place people and the planet at the centre" (SDG17). While being a laudable goal the reality of global inequalities and marketization poses potent barriers that place profit and power over health gains and resource distribution.

Sustainability within the healthcare curriculum is directly relevant to supporting the education of those who are and will be engaged in addressing wicked problems in healthcare settings. Two aspects are discussed next: a program-level initiative and a pedagogical development to promote the type of change that is needed to look ahead to the future with purpose to address evolving complex problems.

PROGRAM-LEVEL INITIATIVES TO PROMOTE SUSTAINABLE DEVELOPMENT

In the context of wicked problems, merely teaching how to address known problems with previously used methods is deemed unlikely to generate a sustainable solution. This is familiar ground in NHS evolution that has seen a succession of initiatives that are announced, launched implemented, and then superseded by the next new idea. The impact of "insufficient staffing results in increased pressure, stress, burnout, lower job satisfaction and a greater inclination to leave among the workforce" (Royal College of Nursing (RCN), 2013).

A large northern English NHS Trust engaged collaboratively with a university provider in seeking a fresh approach to thinking about wicked problems faced in contemporary healthcare. The resulting integrated care program was aimed at front line staff who were close to the daily challenges of safe and effective care delivery. It comprised a two-year program for registered professionals during which participants would undertake a sequence of rotations into different operational areas. The rationale was to be able to appreciate and assess the nature of challenges faced in different service areas about patient care. Allied to this was an introduction to critical thinking and analysis, so that an enhanced questioning approach was taken toward seeing the context and activity and formulating an analysis of what the challenges were, and why they existed, prior to thinking how they might be addressed. Consideration about how they might be addressed was informed by recourse to a broader horizon of thinking, including collaboration with local councilors, patients and carers, public health officials, and service managers. Other similar activities with post-graduate learners have included inviting an asylum seeker into a class to hear their story and discuss how best their needs could be addressed. This was followed by feedback from the asylum seeker to the professionals about the perceived appropriateness of the recommendations themselves. The learning point being to unpick the complexity of a specific case study. Such learning experiences illustrate the need not to impose solutions before exploring and evaluating the understanding and identification of problems and challenges. This in turn directs critical attention back to a shared understanding of what the issues are that are being faced. In a similar way to understanding the complexity of an individual's case there is understanding a service-level case example.

The integrated care program therefore addresses the question: "why are we facing the current issues in this particular setting?" and guides learners to explore and construct an origin story to account for a trajectory of development that has resulted in the current position. This draws on their critical and analytical thinking skills using a suitable model as a guide (such as the PESTLE model to tease out political economic, sociological, technological, legal, and ethical factors). At this stage, political ideology and associated health policies that have been published over several decades are examined to interpret what gave rise to a given policy, how the political narrative made a case of identified national needs that required addressing and how that cast previous developments, to some extent, insufficient to address the challenges of the day. The learning derived is aimed at facilitating students to think beyond given narratives to determine for themselves the "what" and "why" of identified current challenges. The next step in the learning journey is to triangulate with other evidence published about other national healthcare systems and the challenges that they face. Through this parallels are seen in other national healthcare systems that, in a globally connected world, lead to an appreciation of what extrinsically shapes the national challenges. The implication being that any national solution has to include how to work in collaboration with global partners to mediate the extrinsic factors that shape the national situation.

The stage of realization that local action requires a global mind-set (think globally act locally) draws upon a further feature of the program – that of leadership. The NHS has often defaulted to managerialism to enact policy and contain costs, but this has been inadequate and has led to a leadership vacuum and merely serves to perpetuate what already exists. Indeed "leadership" is one of the most influential factors in shaping an organizational culture. Ensuring the necessary leadership behaviors, strategies, and qualities, which are developed, is fundamental (NHS England and Health Education England, 2018). Leadership education includes a range of leadership theories to offer insights into ways in which different people in different settings and at different periods of social history have been influenced by the prevalent idea of the day to move forward to address challenges. Coupled with this is a theme of personal reflection – seeing oneself at a period in time with preconceived ideas about the world and the challenges faced by health services.

The application of leadership learning to real world practice is developed through a reflective analysis of case examples that show how multiagency working requires negotiation skills to navigate a changing landscape of provision. This is a landscape in which participants have conflicting views and even values where common ground needs to be found. This ground is a place where sustainable healthcare provision has to be achieved in the context of finite resources, increasing demand, and workforce deficits. It is not an easy problem to address. One aspect of this is to re-vision existing roles that have developed historically but which might not be agile enough to work effectively to secure the health gains sought in the new landscape. Once again, learning from the experiences of those in other health systems is valuable to inform approaches to sustainable service delivery. A key feature of sustaining any innovation in a health service is to have a sense of longevity – that is to have seen trends over time and have recognized familiar cycles that impact on services. Allied to this is the need for personal resilience – to endure tough periods of transition, to prevail in having a clear ideology that transcends many temporal transitory changes and maintain a sense of direction and investment in striving for health gains.

Personal learning taken forward through a program such as this one on integrated care includes being a life-long learner. In this respect, the development of critical and analytical skills feeds a curiosity in the real world, and the formulation of a coherent and informed position to contribute to sustainable developments (which contrasts with the typical short-termism political agendas in healthcare). At a micro-level continuing practice development (CPD) that minimizes the time that a professional has to spend away from practice includes accessible e-Learning and Health Education England's (2018) "bite-sized" learning to accelerate and extend opportunities for learning in practice. "e-Learning for healthcare" is a portfolio of learning resources available 24/7 facilitating access to CPD (Health Education England, 2019). Moreover, untethering CPD programs from academic credits (and so university-based provision) and assessments enables programs to be more cost effective and so increases availability to services with finite training budgets. The challenge of economic constraints is evident in the "five year forward view" (NHS, 2014) that stated:

> We can design innovative new care models, but they simply won't become a reality unless we have a workforce with the right numbers, skills, values and behaviours to deliver it.

However, resource limitations and funding cuts form part of the problem (Greatbeach, 2016) which the RCN claims has led to nurses not being able to access necessary CPD (Gilroy, 2018).

Given that integrated care is centered on interdisciplinary working it follows that learning together is central to the necessary cultural shift required for practice (Nancarrow et al., 2013). The UN SDGs focus on partnership working and collaborations, not just with healthcare providers but working collaboratively with other services and with the public and patients. For interdisciplinary healthcare education to be effective, other partnerships outside of health need to be sought in order to go beyond creating public health policy to promote healthy public policy. This move into a political arena is about recognizing the health impact of policy in a range of areas from agriculture to trade and industry as well as striving for equity in problem solving and decision-making. This necessarily includes the formalization of health impact assessments in policy arenas.

PEDAGOGICAL-LEVEL DEVELOPMENT TO PROMOTE SUSTAINABLE DEVELOPMENT: SOCIALLY IMMERSIVE LEARNING

A different way of embedding sustainability into a healthcare curriculum is to develop a pedagogy that aligns to the emerging real world context as opposed to perpetuating a model that addressed historical contexts. The example that follows was devised with the challenge of wicked problems in mind and seeking a way to help learners to be fully engaged in social practices to find ways of addressing such problems.

Existing models of education are often predicated on knowledge being transferred to the learner (through a variety of strategies designed to engage the learner) and then through an acquired skill set applied to address known problems. This embodies an idea of what participants are required to learn and how this is achieved. The pedagogies underpinning education also need revisiting to find one that offers the means of sustainable practice in relation to knowledge acquisition, development, and application to wicked rather than contained and defined problems. A model of "socially immersive learning (SIL)" was conceived based on the progression from social behavior to skills and knowledge, rather than the more traditional progression from knowledge and skills to behavior (Gurbutt, Smith, Gurbutt, Duckworth, & Partington, 2019). This recognizes the social shift that has occurred through mobile technology and has changed behavior in terms of the opportunity to communicate and network with others across national and geographical boundaries. This opportunity gives rise to an arena where people can freely choose to start their own networks and draw other interested parties together over issues of common interest. The great advantage of this is the discussion that can occur to explore and examine an issue and appreciate different origin stories, interpretations, and potential areas of agreement and convergence. This type of social network introduces the notion of the global citizen, able to find common values and grounds for collaboration and sharing with people who may never physically meet face to face. The learning here is in harnessing what people are already freely doing – using accessible technology to broaden their reach and horizon. This is the social in SIL in which the participants are engaged and "present" around a shared issue. This contrasts with social networking that remains at a level of consumerist entertainment and diversional activities. The next stage is to identify the skills needed to create or acquire the information needed to conceptualize and define the wicked problem that is being faced. This is where peer learning can occur in a network and where relationships can be strengthened. Participants are globally engaged with altruistic motives to be net contributors rather than net takers. The final dimension of SIL is in the acquisition and even co-creation of knowledge that can be used to inform the account of what the problem is, how and why it has arisen, and what approaches can be taken to address it. This differs in a globally sustainable community from knowledge and expertise being monetarized and only accessible via specific institutions and commercial operations.

SIL supports sustainability because of the fluid nature of networks and the cross-boundary collaboration that rises above the historical definitions of problems and prescribed solutions. It recognizes and accentuates the relational aspects of learning and fits well with the requirement for interdisciplinary and collaborative working as a part of healthcare education as described by the healthcare professional bodies, such as the General Medical Council (GMC, 2018).

CONCLUSION

In conclusion, innovations to develop sustainable curriculum and pedagogy have been developed to address the problems of the present day and anticipate those of the future as opposed to historical ones. The curriculum innovation for sustainable development required participants to critically understand the political, economic, social technological, and ethical contexts driving service redesign but more importantly to critically handle information to formulate an informed position on how to navigate the changing health and social care landscape. A leadership dimension sought to utilize critical thinking and analysis to formulate integrated responses through methods of effective working in a multiagency and professional context. Promoting sustainability thus veered away from accepting short-termism in favor of longer-term interventions that equipped participants with skills of agility to address wicked problems and anticipate situations yet to exist and recognize the need for roles yet to be developed. Multiagency working also must include patient voices so as to promote solutions that maintain the focus on services being for people to gain "a better and more sustainable future for all." Additionally, collaborative and accessible flexible education supports sustainability in terms of continuing practice development to keep abreast of new trends in healthcare. The great endeavor of sustainable development requires a global mind-set to act locally in a meaningful way. This mind-set is argued and can be facilitated through integrated care education promoting critical thinking and SIL.

REFERENCES

British Medical Association (BMA). (2018, December 7). European working hours' time directive. Retrieved from https://www.bma.org.uk/advice/employment/working-hours/ewtd

Frances, R. (2013, February 6). Report of the Mid Staffordshire NHS Foundation Trust Public Inquiry. Retrieved from https://www.gov.uk/government/publications/report-of-the-mid-staffordshire-nhs-foundation-trust-public-inquiry

Gawande, A. (2014). *Being mortal, medicine and what matters in the end.* New York, NY: Metropolitan Book.

General Medical Council (GMC). (2018). Outcomes for graduates. Retrieved from https://www.gmc-uk.org/education/standards-guidance-and-curricula/standards-and-outcomes/outcomes-for-graduates

Gilroy, R. (2018, May). The NHS has denied nurses clinical training, says RCN. Independent nurse. Retrieved from http://www.independentnurse.co.uk/news/the-nhs-has-denied-nurses-clinical-training-according-to-the-rcn/174043/

Greatbeach, D. (2016, September). A false economy; cuts to funding for continuing professional development for nursing, midwifery and allied health. Council of Deans of Health. Retrieved from https://councilofdeans.org.uk/wp-content/uploads/2016/09/19092016-A-False-Economy-CPD-cuts-in-England-2016-17-.pdf

Gurbutt, R., Smith, B., Gurbutt, D. J., Duckworth J. E., & Partington, H. (2019). Socially immersive learning: A new pedagogy. doi:10.21125/inted.2019.0181

Health Education England. (2018) Bitesized teaching. Retrieved from https://www.hee.nhs.uk/our-work/mental-health/bitesized-teaching

Health Education England. (2019) E-learn for healthcare. Retrieved from https://www.e-lfh.org.uk/about/

Hornbeek, J. A., & Peters, B. G. (2017). Understanding policy problems: A refinement of past work. *Policy and Society*, *36*(3), 365–384. doi:10.1080/14494035.2017.1361631

Moberley, T. (2018, January 27). Number of women entering medical school rises. Retrieved from https://www.bmj.com/bmj/section-pdf/959692?path=/bmj/360/8138/Careers.full.pdf

Nancarrow, S. A., Booth, A., Ariss, S., Smith, T., Enderby, P., & Roots, A. (2013). Ten principles of good interdisciplinary team work *Human Resources for Health*, *11*(1), 1. doi:10.1186/1478-4491-11-19

Neivas, I., & Thaver, D. (2015). Work life balance a different scale for doctors? *Frontiers in Pediatrics*, *3*, 115. doi:10.3389/fped.2015.00115

NHS England. (2014, October). Five year forward view. Retrieved from https://www.england.nhs.uk/wp-content/uploads/2014/10/5yfv-web pdf

NHS England. (2015). Integrated primary and acute care systems vanguards. Retrieved from https://www.england.nhs.uk/new-care-models/about/primary-acute-sites/

NHS England. (2016). Integrated care systems. Retrieved from https://www.england.nhs.uk/integratedcare/integrated-care-systems/

NHS England. (2019). Long term plan. Retrieved from https://www.england.nhs.uk/long-term-plan/

NHS England and Health Education England. (2018, March). Integrated urgent care/NHS 111 workforce blueprint, leadership development. Retrieved from https://www.england.nhs.uk/wp-content/uploads/2018/03/leadership-development.pdf

Office for National Statistics (ONS). (2018a, November 6). Earnings and hours worked, UK region, public and private sectors by SOC (1 to 2 digit) 2010 to 2018. Retrieved from https://www.ons.gov.uk/employmentandlabourmarket/peopleinwork/earningsandworkinghours/adhocs/009257earningsandhoursworkedukregionpublicandprivatesectorsbysoc1to2digit2010to2018

Office for National Statistics (ONS). (2018b, September 9). How has life expectancy changed over time? Retrieved from https://www.ons.gov.uk/peoplepopulationandcommunity/birthsdeathsandmarriages/lifeexpectancies/articles/howhaslifeexpectancychangedovertime/2015-09-09

Royal College of Nursing (RCN). (2013, April). Beyond breaking point? A survey report of RCN members on health, wellbeing and stress. Retrieved from https://www.rcn.org.uk/-/media/royal-college-of-nursing/documents/publications/2013/april/pub-004448.pdf

Royal College of Physicians. (2005, December). Doctors in society: Medical professionalism in a changing world. Report of a Working Party. Retrieved from https://cdn.shopify.com/s/files/1/0924/4392/files/doctors_in_society_reportweb.pdf?15745311214883953343

The National Archives. (1942, November 25). Beveridge report: Social insurances and allied services. Retrieved from http://filestore.nationalarchives.gov.uk/pdfs/small/cab-66-31-wp-42-547-27.pdf

United Nations. (2018). The sustainable development goals report 2018. Retrieved from https://unstats.un.org/sdgs/report/2018/overview/

World Health Organization (WHO). (2015). World report on ageing and health. Retrieved from https://www.who.int/ageing/publications/world-report-2015/en/

World Health Organization (WHO). (2017). Models of health care provision: Guidance tools. Retrieved from http://www.who.int/hac/techguidance/tools/disrupted_sectors/module_07/en/index3.html

CHAPTER 2

EMBEDDING SUSTAINABLE DEVELOPMENT IN THE CURRICULA: LEARNING ABOUT SUSTAINABLE DEVELOPMENT AS A MEANS TO DEVELOP SELF-AWARENESS

Louise Manning and Luis Kluwe de Aguiar

ABSTRACT

Sustainable development is generally a difficult concept to explore in an educational setting. Those attempting to teach it require the inter- and trans-disciplinary skills to be able to foster in the students not only the knowledge assimilation but also the deep acquisition of values (Faham, Rezvanfar, Mohammadi, & Nohooji, 2017; Lambrechts, Mulà, Ceulemans, Molderez, & Gaeremynck, 2013). Sustainable education is embedded in the curriculum of Food Technology students throughout their four-year degree at Harper Adams University. In this chapter, a brief description of how the curricula are organized is presented. Emphasis is given to one module in the final year of the degree where the content and pedagogical practice as well as the pedagogic nature of the learning environment is analyzed in the light of the literature. The teaching and learning practices developed, and the extent to which they equip the students with skills such as critical thinking, analysis, reflection, and complex problem solving is explored. Two case studies are presented, namely the Loess Plateau (China) and the Chiapas Coffee Farmers (Mexico). These serve to

Integrating Sustainable Development into the Curriculum
Innovations in Higher Education Teaching and Learning, Volume 18, 25–39

ISSN: 2055-3641/doi:10.1108/S2055-364120200000018020

contextualize the extent that students can draw from teaching and learning strategies in the module as well as other experiences (personal or from other modules) and use these to frame the students' learning.

Keywords: Sustainable education; problem solving, case study learning; curricula design; critical thinking; self-awareness; reflection

INTRODUCTION

The World Commission on Environment and Development defines sustainable development as "development that meets the needs of the present without compromising the ability of future generations to meet their own needs" (World Commission on Environment and Development (WCED), 1987). Sustainable development is a multifaceted, multidisciplinary, dynamic, and intertemporal concept encompassing social, ethical, and political behavior (Al Rawahy, 2013, p. 400). Thus, it is a difficult concept to explore in educational settings. Sustainable development can be described simply as an abstract concept. However, efforts have been made to make the term more tangible. Indeed, Elkington (1994) sought to measure sustainability through a metrics- and accounting-based approach that built on the traditional "economic" bottom line, that is, it built upon our understanding of measuring financial performance to then develop the "triple bottom line." This tool is a means to demonstrate the delivery of sustainability goals in terms of economic (financial), environmental, and social performance.

The United Nations Educational, Scientific and Cultural Organization (UNESCO, 2007, p. 6) stated that:

> Education has the potential to play a major role in the future realisation of a vision of sustainability that links economic well-being with respect for cultural diversity, the Earth and its resources.

Therefore, universities have a role in educating individuals and developing knowledge that can be transformed into social skills (human and social capital). Universities also have a part to play in driving governance at a national and regional level where they can provide expertise, knowledge, develop holistic networks and as a result contribute to regional development via increased physical and financial capital (Al Rawahy, 2013; Sedlacek, 2013). Thus, a question arises as to whether sustainability should be taught as an individual topic or instead be embedded holistically across the curricula (Lambrechts et al., 2013; Lazzarini, Pérez-Foguet, & Boni, 2018). The first approach of identifying sustainability as an element of a curriculum matrix works within ridged, focused often siloed learning structures that deliver content-based learning and knowledge transfer. However, this style of learning will not allow students to consider their own interaction with the notions of sustainability and then to develop their own learning program where they can explore their own attitudes, knowledge skills, and values in order to become change agents in their future work roles (Lambrechts et al., 2013).

Furthermore, the United Nations (UN) declaration of the Decade of Education for Sustainable Development (DESD) by UNESCO gave attention to the development of a holistic, inter- and transdisciplinary approach to embedding sustainable development into the curricula through developing sustainable education (SE) or "education for sustainable development" (ESD). This means that there is attention paid to acquiring values rather than just knowledge assimilation (Faham et al., 2017; Lambrechts et al., 2013; UNESCO, 2007, 2009). The pedagogic nature of the learning environment is important. The ability to reflect on the complex problems associated with sustainability issues requires the development of student-focused curricula that promote active learning and develop techniques that include problem solving and critical reflection, case studies, discussions, workshops, and problem-based learning (Figueiró & Raufflet, 2015). The Organisation for Economic Co-operation and Development (UNESCO, 2009) states that there are two themes to ESD. Firstly, an emphasis on the students' gaining an understanding of both the causes and the resultant impacts of key SD issues, including their economic, environmental, social, and cultural interlinkages. Secondly, concentration on the capacity development at the community and individual level helps to mitigate these challenges. UNESCO (2009) also lists 14 learning outcomes in terms of key skills and competences that are of interest here (see Table 1), which have been grouped thematically in terms of adaptability, change and crisis management, relationship building, and high-level thinking. In their explanation, competence is "the ability to successfully meet complex demands in a particular context through the mobilization of psychological prerequisites including both cognitive and non-cognitive aspects" (Rychen & Salganik, 2003, p. 43). Therefore, a transition to competence-based education approaches first requires the program to center on the level of competence (skills,

Table 1. Sustainable Development: Key Learning Outcomes – Skills and Competences.

Adaptability
Applying learning in a variety of life-wide contexts
Change and crisis management
Decision-making, including uncertain situations
Dealing with crises and risks
Relationship building
Acting with responsibility locally and globally
Acting with respect for others
Ability to identify and clarify values
Participating in democratic decision-making
Negotiating and consensus building
Identifying stakeholders and their interests
High-level thinking
Critical reflective thinking
Understanding complexity/systemic thinking
Futures thinking
Understanding interrelationships across disciplines (interdisciplinary thinking)

Source: Adapted from UNESCO (2009).

knowledge, ability to apply knowledge, etc.) that should be achieved and then to see how this is assessed (Lambrechts et al., 2013).

Others have also identified key competences essential in ESD and these have been thematically grouped (Table 2) in line with the categories in Table 1. Wiek et al. (2015) built on this and have suggested five key competencies related to SE:

1. Systems thinking is a subset of higher-level thinking as described in Tables 1 and 2.
2. Interpersonal competencies.
3. Normative competences are elements of the group of relationship building competencies described in other literature.
4. Anticipatory competences relate to adaptability as described above.
5. Strategic competencies relate to the iterative category here of change and crisis management.

These competences then need to be considered when developing a learning program. Three key elements are essential when developing a module, course, or program: first, defining learning outcomes; second, indicating how the course will be delivered; and, third, the method(s) of assessment (Lozano, Ceulemans, &

Table 2. Competences of Interest in Developing Educational Programs That Focus on Sustainable Development.

Adaptability
International focus
Attitude
Change and crisis management
Action skills
Competence in planning and implementation skills
Relationship building
Commitment
Competence in participatory skills
Competence in self-motivation and motivating others
Emotional intelligence
Empathy compassion and solidarity
Respect
Responsibility
Transcultural understanding
Values and ethics
High-level thinking
Competence in cosmopolitan perception, transcultural understanding, and co-operation (understanding complexity/systems thinking)
Competence in distanced reflection on individual and cultural models (understanding complexity/ systems thinking)
Future/foresighted thinking
Interdisciplinary thinking
Self-reflection and peer assessment

Source: Adapted from Muijen (2004), De Haan (2006), Sleurs (2008), Roorda (2010), and Lambrechts et al. (2013).

Seatter, 2015). There are some distinct characteristics of teaching, learning, and assessment methods that are crucial for developing competences in education that center on interactive and participative methods, action-oriented methods, research methods, methods for evaluation and reflection (Table 3).

Education focused on sustainable development requires complex systems thinking (as shown in Tables 1 and 2, and this influences educational approaches and practice that reflect the multiplicity of acceptable answers rather than singular response to a given question (Du et al., 2013). Interdisciplinarity is also an important prerequisite for problem solving. Bearing in mind the holistic nature of ESD, the learning outcomes developed are rarely addressed by one single academic discipline as the study area can include geography, nutrition, environmental science, biology, agriculture, political science, and others (Annan-Diab & Molinari, 2017). Furthermore, Annan-Diab and Molinar (2017) argue are not only the disciplines of academia interconnected in ESD but also the issues that need to be addressed which means that deep complexity lies at the heart of the problem. A challenge within the learning setting is the teachers lack this interdisciplinarity, knowledge, and understanding as unless teaching programs can be co-taught or co-created with others including students, it will be a barrier to embedding ESD (Friman et al., 2018).

Wicked food policies are characterized by

> the high complexity of their different elements, subsystems, and interdependencies between them; an extraordinary uncertainty regarding the prevailing risks, the consequences of actions addressing them, and constantly changing patterns; and a high divergence and fragmentation of the viewpoints, values and intentions of the actors involved. (Head, 2008, p. 103)

It is important within SE that wicked policies are introduced, considered, and critiqued as part of the curricula development and implementation process.

Table 3. Characteristics of Teaching, Learning, and Assessment Methods for Developing Competences in Education.

Characteristics	Teaching, Learning, and Assessment Methods
Interactive and participative methods	Socrates method, group discussion, role play, group or personal diary, brainstorming, peer assessment
Action-oriented methods	Learning through internships, solving real-life problems, authentic case studies. Project-based learning
Research methods	Bibliographic research, problem analysis, value clarification, case studies, concept mapping, etc.
Self-evaluation, group evaluation, tutor evaluation	Self-assessment, peer assessment, *feeding up*, for example, the use of exemplars; *feeding back* by tutor on formative and summative assessments, *feeding forward* by the tutor after assessment
Reflection	*Content reflection* via examining content or evidence, *process reflection*, that is, checking on the problem-solving strategies and *premise reflection* when the problem itself, for example, climate change is challenged

Source: Adapted from Lambrechts, Van der Haute and Vanhoren (2009), Lambrechts et al. (2013), Du, Su, and Liu (2013), Leal Filho et al. (2018).

Furthermore, the pedagogic elements outlined in Table 3 suggest a more democratized form of learning. Historically, education programs have been prescriptive rather than co-created between the tutors and the learners. Prescriptive education is based on a "monopoly of production and control of knowledge" by the knowledge owner (Nasibulina, 2015), whereas active learning and the associated co-creation of a curriculum (Table 4) lead to better skills development and competency within the student body, professional achievement, student retention, improved course approval, and greater sense of ownership and a sense of belonging (Nielsen & Yahya, 2013). This means that the learning environment needs to change from a closed process of teacher-centered teaching to an open process focused on student development (Barth & Rieckmann, 2012; Kolb, Fröhlich, & Schmidpeter, 2017). A student-centered learning environment includes features of both the responsibility and the self-management and forms of collective learning as all social actors are involved in the learning experience (Caniglia et al., 2018). Social learning occurs when "divergent interests, norms, values and constructions of reality meet in an environment that is conducive to learning" (Wals & van der Leij, 2007, p. 18).

Social learning in groups is not planned and prescribed, instead it is emergent and arises from the interaction between actors with four elements which impact on learning: content knowledge, personal development, teaching competence, and processual learning, that is, the facilitation of social learning (Cincera et al., 2018). Thus, social learning forms a key element of ESD.

In summary, ESD is a competency-based learning experience where students should develop over the course of the learning experience through gaining knowledge, developing personally and as a result be more able to solve the problem and consider challenging cross-disciplinary issues. The next section of the book chapter looks more specifically at educational case studies and how they have been developed to increase social learning and a student-centered experience.

Table 4. Examples of Tutor (Lecturer) and Co-created Curricula Content.

Tutor (Lecturer) and Cocreated Curricula Content
Tutor (lecturer) created
Curricula handbook detailing curricula brief, specifications, guidelines, teaching plans, assessment details, etc.
Knowledge objects that support learning
Topical definitions linking topical learning activities and related assessments
Communications channel for lecturer-to-student, student-to-lecturer, and student peers
Internet and multimedia links as a form of learning resource
Create relevant and appropriate links from specific topical contents to additional resources to encourage students
Co-created
Interactions and learning engagement
Shared media and information
Co-created content in fora and learning spaces

Source: Adapted from Nielsen and Yahya (2013).

HARPER ADAMS UNIVERSITY: THE TEACHING OF SUSTAINABLE DEVELOPMENT IN A FOOD TECHNOLOGY CURRICULUM

The teaching of sustainability in a food technology degree is considered to be integral to the curriculum and not an add-on. Traditionally, food technology curricula attempted to equip future professionals with skills essential for their technical field of expertise. However, career progression entails food professionals becoming managers of operations and processes, and this requires a wider set of skills and competencies than the narrow focus of physical–chemical and biotechnology disciplines.

Furthermore, increasingly, food professionals are pressured not only to maximize resource use by delivering yield improvement and economic profit, but also to address the delivery of the triple bottom line. Schaltegger, Lüdeke-Freund, and Hansen (2012) propose that there is a business case for sustainability that aims to realize economic return through environmental and social activities. In order to provide the students with an understanding of the complexity of these factors and at the same time allow them to make connections between theory, practice, and work experience so that the teaching and learning of topics becomes more meaningful, the sustainability curricula was constructed as proposed in Fig. 1.

In Year 1, students consider general issues affecting the food sector in a module "Issues in Global Food Production." The sessions are organized around a logical sequence of current food industry topics over 23 weeks. As an introductory module, the students build knowledge and the group converge toward a common cognition level. Year 1 students are joining the degree with a range of specialisms stretching from biology, geography, chemistry, sports science, etc. Therefore, their previous knowledge on sustainability is not uniform meaning that for some the concepts covered in the module are being addressed for first time, while for others the sessions might be a form of revision. Yet, to progress to Year 2, the group needs to achieve a shared level of knowledge. Bloom (1956) defines three "domains" of learning cognitive (knowing); conative or psychomotor (doing) and affective (feeling). He suggests that cognitive learning encompasses at the core the ability to undertake simple knowledge recall and then to develop the intellectual skills of knowledge, comprehension, application, analysis, synthesis, and evaluation. Throughout the module, the students attempt a range of formative quizzes that check their level of understanding and allow them to apply their knowledge

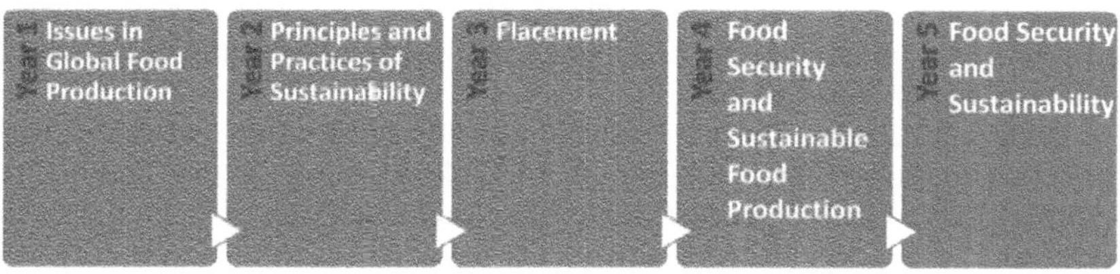

Fig. 1. Sustainability Route.

in preparation for the summative assessment. At the end of the module, they are tested through the means of a formal written examination which is 100% of the overall course mark.

In Year 2, the students join a module entitled Principles and Practices of Sustainability. The module was designed to provide a learning environment that encourages learners to connect their learning from Year 1, apply the principles, and reflect on complex sustainability issues. The teaching and learning strategy is built around a student-focused curricula that promotes active learning, problem-solving, and problem-based learning (Figueiró & Raufflet, 2015). As potential future managers, during the first half of the module, emphasis is given to developing student awareness of the underlying sustainability demands around reuse and recycling of materials such as water, energy, waste, and human impact on the planet. The term "Anthropocene" is used for highlighting the pivotal role since the late eighteenth century of humans and their activities in framing geology and ecology through increasing the emissions of greenhouse gases such as carbon dioxide (CO_2) and methane (CH_4) (Crutzen & Stoermer, 2000). Some suggest a solution for ameliorating the detrimental impact of food production, for example, with the application of sustainable intensification (Rockström et al., 2017) methods or mechanisms to reduce water stress (Rockström & Karlberg, 2010; Sivapalan et al., 2014), and these are considered. The second half of the module considers alternative protein sources (insects, lab-grown meat, pulses), and the implication of food manufacturers exercising sustainable choices when purchasing food ingredients. The length of teaching is 21 weeks. The topics are organized in themes that last three to four weeks allowing students to construct more meaningful experiences as they consider particular issues (Bruner, 1960).

A spiral curriculum has "an iterative revisiting of topics, subjects or themes throughout the course" (Harden & Stamper, 1999, p. 141). Harden and Stamper (1999) argue that as topics are revisited the level of difficulty in terms of knowledge, skills, or demonstrable competencies increases, and the role of the teacher is to relate new learning to previous activities so ultimately there is student progression. Thus, it is important to highlight to students that Year 2 serves as a preparation for Year 3 working in the food industry and putting the theory into practice as well as the skills and competencies they have developed. During the Placement year (a minimum of 44 weeks), some students will not only be seeing sustainability practices on a daily basis but also will have the opportunity to write a formally assessed project report for the company they are working for. This usually consists of solving a problem or improving some procedures including a focus on sustainability. This helps students to understand cause–effect relationships and likely consequences at company, community, and individual levels as mentioned by the UNESCO (2009).

In Year 4, the students are enrolled in the Food Security and Sustainable Food Production module, with a duration of 15 weeks learning. In that module, the emphasis is on the fundamental drivers for food insecurity, but with a strong component on sustainable development. The teaching content is positioned from a food systems' approach that enables students to solidify their understanding of contextual complexity and the need to "sophisticate" their critical analysis from a bi-dimensional linear thinking approach to systems-based thinking. The students are exposed to key case studies (see below) so that their ability to evaluate and critically reflect on sustainability issues develops further. Indeed, a case study based approach rejects the notion of the teacher as information provider and, instead, allows students to develop their own knowledge and experience through deeper learning (Steiner & Posch, 2006).

The summative assessment requires the students to propose solutions for food insecure families in the UK with particular emphasis on using food that would otherwise be waste. Using the British Nutrition Foundation ingredient calculator, students need to analyze the nutritional content of the food waste used and develop or adapt an existing recipe so that it is simple and cost effective for their target consumers. The students then need to prepare and assess the final product organoleptically. All recipes are collated into a repository of ideas and solutions for household food waste. This module marks the end of the undergraduate program.

For those students who continue onto a Food Industry Management MSc. course at the university, the module Food Security and Sustainability is designed to develop their knowledge and skills further. There are multiple access routes to the Masters Course, either directly from the undergraduate program or via a work-based route. Therefore, it is very important to design the module to allow students from different learning backgrounds and types of work experience to have the opportunity not only to be acquainted with some of the topics, but also to consider sustainability and sustainable development in the context of their own and the collective experiences. This mutual activity may also act as a means to focus or even change their career path. The themes covered in the undergraduate program are revisited, but the students are required to approach the topics with more in-depth analysis and evaluation. Emphasis is given to alternative solutions to sustainable development focusing more on the urban environment. The notion of food systems is looked at in more depth in the module in the light of the extent to which food technology can be used in food production particularly regarding urban agriculture, vertical farming, rooftop, hydroponics, and aquaponics. These technological approaches of sustainable intensification are compared and contrasted with other forms of agriculture, such as conventional and organic/eco-agriculture. Additional case studies are used in this module that are new to the students in terms of course content and have relevancy in the context of the human impact on the planet in terms of planetary boundaries and tipping points. This enables the students to apply theory to actual issues. Two case studies of these are now explored in more detail as they provide an opportunity in this chapter for reflection on how such case studies can be used in other curricula.

CASE STUDIES

The Loess Plateau (China)

This case study is published by the World Bank "Restoring China's Loess Plateau" (which is supported by a video and report, see World Bank, 2007). Other source materials are available online that can be used by students. The case study itself frames an international context and the interaction between a province in China and an international institution, the World Bank. As a result, the students should be intellectually equipped to understand the viewpoints of an international organization as well as a set of transcultural challenges. Students should be able to accept values and ethics distinct from their own. This understanding of cultural relativity is critical within the learning process. This relies on the lecturer's ability to adapt the case to interface with their own knowledge and experiences and use these to give the students additional context. The Loess Plateau case study depicts a generalized environmental breakdown caused by human activity over many centuries that is followed by its subsequent partial recovery through policy intervention to enable landscape recovery. The Loess Plateau and its recovery is emblematic of not only the detrimental impact of human activity on the natural environment, but also the impact of extreme food insecurity. The case allows the lecturers to explore many other lines of enquiry that will develop the students' knowledge around themes such as deforestation, soil erosion, endangering of biodiversity, water quality, crop failure, use of agrochemicals abandoned children, schooling, malnutrition, information use, and use of policy interventions, among others. However, developing these themes depends on the lecturer's individual level of multidisciplinary knowledge and that of the whole teaching team. Discussion in class allows the students to embed their knowledge through exploring these case studies and sharing their ideas. Being able to articulate arguments verbally in this group context is a key learning opportunity to then being able to construct written arguments with supporting evidence within their summative assessments. To prepare for the case study, three teaching weeks are used to consider differences in agricultural production methods. The students become acquainted with conventional, organic, intensive methods of food production and the role of the Green Revolution and its policy aim of alleviating food insecurity. This is felt necessary to help with a better contextualization of some of the issues tackled in the case study. Further framing of the history of China in the latter part of the twentieth century is provided so that the students can appreciate the process of the Chinese society transitioning from a centrally planned socialist economy toward a centrally planned market economy. It is important that the students have a holistic view of the social–political–economic processes that have taken place in China in recent decades. Therefore, a multidisciplinary focus on the historical, socioeconomic, and policy dimensions should serve as the framework for the unpacking of the case to allow for a deeper student learning experience. After the preparatory lectures, the Loess Plateau case is presented and at the end of the session a handout is given with a summary of the sustainable development activities and achievements. Thereafter, the students have to read around the topic and are asked to watch the video produced by the World Bank in their own time. When the students return to the classroom, each has to present one aspect of the case to the rest of the group and then justify why that was an important learning point. This supports the development of their communication and critical analysis skills. It is natural that with food technology students, the issues around food insecurity, nutrition, and use of resources to produce food tend to be highlighted first. However, other themes emerge and the conversation allows a discussion on systems-based thinking The teacher facilitates this process supporting knowledge exchange between students. Finally, the students are encouraged to discuss and critique the extent to which the Loess Plateau Rejuvenation project has affected the livelihoods of the local population. The environmental degradation and its detrimental impact not only felt on the Loess population but also farther afield is a main learning point as it highlights the fragility of ecosystems if policy makers and society fail so see what problems can occur and the need for a holistic process to find solutions in a systems-based approach.

From Tables 1 and 2, the proposed key learning outcomes such as adaptability, change and crisis management, relationship building, and high-level thinking are addressed in this case study. Yet, the challenge to deliver these learning opportunities depends on the lecturers' ability to facilitate learning in a manner that means that the students can establish relationships based on the issues discussed, prior knowledge, and their personal experience of and interaction with those topics in other contexts. For some students, on a typical Food Technology degree where half of the disciplines covered are considered to be "hard science," it is a hurdle to make them understand how the so called "soft science" disciplines can explore multiple dimensions and still provide valid solutions. Yet, to make sure the intended learning outcomes are achieved, the lecturer has to demonstrate the depth of this case with all its systemic complexities and far reaching implications. The case study itself presents the reader with a crisis situation. This allows the student to reflect on a set of cause–effect conditions that create both positive and negative externalities. Therefore, their ability to identify the main problem, clarify the priorities and the likely counter effect on the many stakeholders involved is honed and developed. Indeed, the lecturer should avoid using the Loess Plateau Rejuvenation project case in a shallow bi-dimensional way and instead facilitate learning that focuses on relationship building and critical reflective thinking is encouraged. As part of the Year 4 curricula, the students studying in parallel to this module other disciplines such as supply chain management and food ethics. Thus, by the point during the term when this case is presented, the students ought to be able to apply their interdisciplinary cross-curricula thinking skills. The students should be also competent enough to both evaluate in depth and relate to the ethics of sustainability in the two situations presented to them.

After the Harvest; Fighting Hunger in the Coffee Lands

This case study focuses on the Chiapas Coffee Farmers in Mexico and highlights a different dimension of sustainable development in agricultural communities in North America. Despite upland farmers having opted to produce coffee beans to meet the demand within the niche Fair Trade market, their embedded challenges with household food security means that their family members are subject to periods of malnutrition which have become endemic in the mountains of Mexico. Lured by higher market value for their "cash crop" the farmers fail to set land aside for a vegetable garden or some domestic animals that would see them and their families through the lean months of financial income. A video produced by the Non-Governmental Organization (NGO) "After the Harvest" is watched during the lecture as a means to encourage the students to discuss the points raised on food insecurity (After the Harvest, 2011). Similar to the previous case, an understanding of the different agricultural production systems that could be available in this case study setting is important for the students. This information is provided in the preparatory lectures to help the students to better contextualize their analysis. As some of the dialogue is in Spanish, there are many nuances that the students might miss in terms of the detailed meaning in the conversations in the video. This gap has to be supported in the sessions by the lecturer. Again the lecturer's multidisciplinary preparedness is vital to lead students usually constrained by a positivistic and narrow technology-based content to draw from previous learning experiences so they can understand the wider context of the challenges for these families. This means they tend to focus first on the problem of malnutrition and hidden hunger (vitamin and mineral deficiency, stunting, wasting). The students are encouraged to watch the documentary a second time to help them to understand the complexities of rural migration, and the impact that the lack of money/credit has on the mountain population during the lean months. Furthermore, the documentary also helps to point out the juxtaposition of fairness and "unfairness" of a coffee "fair trade" system that focuses on the use of natural resource to underpin the exporting of the coffee beans, thus reducing land availability for their own food production. This wider issue of resource grabbing from impoverished communities to supply a market demand in more economically developed countries is an important theme for the students to understand. There is also the aspect of external intervention in this case study from NGOs trying to ameliorate the situation of the farmers and their families' situation by teaching basic animal husbandry techniques and encouraging them to keep vegetable gardens to supplement their diet.

The Chiapas Coffee Farmers' case study provides the students with a good opportunity for contextualized reflection. The ability to critically reflect is a key aspect of a graduate's development, and this will be formally assessed in their assignments. By exposing the students to the case study some then feel ready to share past learning experiences. Often they bring examples and experiences from when they worked as volunteers during a Gap Year. Therefore, as proposed by Nielsen and Yahya (2013), without, the students realizing they are engaging in the co-creation of learning content.

SUMMARY

Both these examples highlight to the students the notions of wicked policies that are complex and contain subsystems and interdependencies (Head, 2008). The contextualization of risks and benefits also means that students need to be confident to consider and assess uncertainty as a concept. Learning environments that promote experiential learning give rise to emotional situations as students test themselves in terms of their knowledge and their skills, and this is especially so when there is a lack of certainty on what is the appropriate answer is. Lackéus (2014) states that teaching without the use of emotion, risks getting lost in "boredom and stagnation." Emotional intelligence drives certain emotions, perceptions, the ability to assimilate, and in turn critical thinking and problem solving; therefore, embedding in the curricula opportunities to develop emotional intelligence is a critical element of the student experience (Jordan, Ashkanasy, Hartel, & Hooper, 2002).

CONCLUSIONS

This chapter has considered how spiral curricula can be developed within an SE context. The learning outcomes suggested for sustainability-related subjects (OECD, 2009), and the works of Lambrechts et al. (2013) among others have been used to reflect on the extent to which the teaching and learning practices employed have developed competences and skills in the students studying the course. These skills include critical thinking, analysis, reflection, and complex problem solving. Two case studies are presented in the chapter: the Loess Plateau example (China) and the Chiapas Coffee Farmers (Mexico). These two case studies provide a rich backdrop and serve to contextualize the extent to which the students can consider both positivist and interpretivist mind-sets in approaching each example. Furthermore, particularly, the analysis of the Chiapas Coffee Farmers' case study enables the students to also explore the ethics of sustainable development and through that lens also establish relationships with topics studied in other modules. This allows students to frame through the use of case studies their problem solving, and they can then co-create knowledge with each other and the lecturer. The use of a case study learning approach combined with problem solving is a key to the pedagogy used in this course. Problem solving using real-life contexts allows the students to gain a deeper meaning of the topics covered and also to work with examples that have aspects of uncertainty and also imperfect levels of information. This approach allows the students to develop a level of self-awareness and identify areas of self and mutual growth within the context of the learning environment.

REFERENCES

After the Harvest. (2011). After the harvest; fighting hunger in the coffee lands. Retrieved from http://aftertheharvestorg.blogspot.com/p/home.html

Al-Rawahy, K. H. (2013). Engineering education and sustainable development: The missing link. *Procedia-Social and Behavioral Sciences*, *102*, 392–401.

Annan-Diab, F., & Molinari, C. (2017). Interdisciplinarity: Practical approach to advancing education for sustainability and for the sustainable development goals. *The International Journal of Management Education*, *15*(2), 73–83.

Barth, M., & Rieckmann, M. (2012). Academic staff development as a catalyst for curriculum change towards education for sustainable development: An output perspective. *Journal of Cleaner Production*, *26*, 28–36.

Bloom, B. S. (1956). *Taxonomy of educational objectives*. New York, NY: Longman.

Bruner, J. S. (1960). *The process of education*. Cambridge, MA: Harvard University Press.

Caniglia, G., John, B., Bellina, L., Lang, D. J., Wiek, A., Cohmer, S., & Laubichler, M. D. (2018). The global curriculum: A model for transnational collaboration in higher education for sustainable development. *Journal of Cleaner Production*, *171*, 368–376.

Cincera, J., Biberhofer, P., Binka, B., Boman, J., Mindt, L., & Rieckmann, M. (2018). Designing a sustainability-driven entrepreneurship curriculum as a social learning process: A case study from an international knowledge alliance project. *Journal of Cleaner Production, 172*, 4357–4366.

Crutzen, P. J., & Stoermer, E. F. (2000). The Anthropocene. *IGBP Newsletter, 41*, 17–18.

De Haan, G. (2006). The BLK '21'programme in Germany: A 'Gestaltungskompetenz'-based model for education for sustainable development. *Environmental Education Research, 12*(1), 19–32.

Du, X., Su, L., & Liu, J. (2013). Developing sustainability curricula using the PBL method in a Chinese context. *Journal of Cleaner Production, 61*, 80–88.

Elkington, J. (1994). Towards the sustainable corporation: Win–win–win business strategies for sustainable development. *California Management Review, 36*(2), 90–100.

Faham, E., Rezvanfar, A., Mohammadi, S. H. M., & Nohooji, M. R. (2017). Using system dynamics to develop education for sustainable development in higher education with the emphasis on the sustainability competencies of students. *Technological Forecasting and Social Change, 123*, 307–326.

Figueiró, P. S., & Raufflet, E. (2015). Sustainability in higher education: A systematic review with focus on management education. *Journal of Cleaner Production, 106*, 22–33.

Friman, M., Schreiber, D., Syrjänen, R., Kokkonen, E., Mutanen, A., & Salminen, J. (2018). Steering sustainable development in higher education: Outcomes from Brazil and Finland. *Journal of Cleaner Production, 186*, 364–372.

Harden, R. M., & Stamper, N. (1999). What is a spiral curriculum? *Medical Teacher, 21*(2), 141–143.

Head, B. W. (2008). Wicked problems in public policy. *Public Policy, 3*(2), 101–118.

Jordan, P. J., Ashkanasy, N. M., Hartel, C. E. J., & Hooper, G. S. (2002). Workgroup emotional intelligence. Scale development and relationship to team process effectiveness and goal focus. *Human Resource Management Review, 12*(1), 195–214.

Kolb, M., Fröhlich, L., & Schmidpeter, R. (2017). Implementing sustainability as the new normal: Responsible management education – From a private business school's perspective. *The International Journal of Management Education, 15*(2), 280–292.

Lackéus, M. (2014). An emotion based approach to assessing entrepreneurial education. *The International Journal of Management Education, 12*(1), 374–396.

Lambrechts, W., Mulà, I., Ceulemans, K., Molderez, I., & Gaeremynck, V. (2013). The integration of competences for sustainable development in higher education: An analysis of bachelor programs in management. *Journal of Cleaner Production, 48*, 65–73.

Lambrechts, W., Van den Haute, H., & Vanhoren, I. (2009). *Duurzaam hoger onderwijs. Appel voor verantwoord onderrichten, onderzoeken en ondernemen. [Sustainable higher education. Appeal for responsible education, research and operations.]* Leuven, Belgium: LannooCampus.

Lazzarini, B., Pérez-Foguet, A., & Boni, A. (2018). Key characteristics of academics promoting sustainable human development within engineering studies. *Journal of Cleaner Production, 188*, 237–252.

Leal Filho, W., Raath, S., Lazzarini, B., Vargas, V. R., de Souza, L., … Orlovic, V. L. (2018). The role of transformation in learning and education for sustainability. *Journal of Cleaner Production, 199*, 286–295.

Lozano, R., Ceulemans, K., & Seatter, C. S. (2015). Teaching organisational change management for sustainability: Designing and delivering a course at the University of Leeds to better prepare future sustainability change agents. *Journal of Cleaner Production, 106*, 205–215.

Muijen, H. (2004). Integrating value education and sustainable development into a Dutch University curriculum. *International Journal of Sustainability in Higher Education, 5*(1), 21–32.

Nasibulina, A. (2015). Education for sustainable development and environmental ethics. *Procedia-Social and Behavioral Sciences, 214*, 1077–1082.

Nielsen, D., & Yahya, Y. (2013). Co-creating and mapping curricula to the VLE. *Procedia Technology, 11*, 710–717.

Rockström, J., & Karlberg, L. (2010). The quadruple squeeze: Defining the safe operating space for freshwater use to achieve a triply green revolution in the Anthropocene. *Ambio, 39*(3), 257–265.

Rockström, J., Williams, J., Daily, G., Noble, A., Matthews, N., … de Fraiture, C. (2017). Sustainable intensification of agriculture for human prosperity and global sustainability. *Ambio, 46*(1), 4–17.

Roorda, N. (2010). *Sailing on the winds of change. The Odyssey to sustainability of the universities of applied sciences in the Netherlands*. Ph.D. thesis, Maastricht University, Maastricht, The Netherlands.

Rychen, D. S., & Salganik, L. H. (Eds.). (2003). *Key competencies for a successful life and well-functioning society*. Seattle, WA: Hogrefe Publishing.

Schaltegger, S., Lüdeke-Freund, F., & Hansen, E. G. (2012). Business cases for sustainability: The role of business model innovation for corporate sustainability. *International Journal of Innovation and Sustainable Development*, *6*(2), 95–119.

Sedlacek, S. (2013). The role of universities in fostering sustainable development at the regional level. *Journal of Cleaner Production*, *48*, 74–84.

Sivapalan, M., Konar, M., Srinivasan, V., Chhatre, A., Wutich, A., … Rodríguez-Iturbe, I. (2014). Socio-hydrology: Use-inspired water sustainability science for the Anthropocene. *Earth's Future*, *2*(4), 225–230.

Sleurs, W. (Ed.). (2008). *Competences for ESD (Education for Sustainable Development) Teachers. A framework to integrate ESD in the curriculum of Teacher Training Institutes (Brussels)* (p. 1-90). Retrieved from http://www.unece.org/fileadmin/DAM/env/esd/inf.meeting.docs/EGonInd/8mtg/CSCT%20Handbook_Extract.pdf.

Steiner, G., & Posch, A. (2006). Higher education for sustainability by means of transdisciplinary case studies: An innovative approach for solving complex, real-world problems. *Journal of Cleaner Production*, *14*(9–11), 877–890.

The United Nations Educational, Scientific and Cultural Organization (UNESCO). (2007). *The UN Decade for Education for Sustainable Development (DESD 2005–2014): The first two years*. Paris, France: UNESCO.

The United Nations Educational, Scientific and Cultural Organization (UNESCO). (2009). Learning for a sustainable world: Review of contexts and structures for education for sustainable development. Section for DESD coordination, Division for the Coordination of United Nations Priorities in Education, UNESCO. Retrieved from www.unesco.org/education/justpublished_desd2009.pdf

Wals, A. E. J., & van der Leij, T. (2007). Introduction. In A. E. J. Wals (Ed.), *Social learning towards a sustainable world. Principles, perspectives, and praxis* (pp. 17–32). Wageningen, The Netherlands: Wageningen Academic Publishers.

Wiek, A., Bernstein, M., Foley, R., Cohen, M., Forrest, N., … Withycombe Keeler, L. (2015). Operationalising competencies in higher education for sustainable development. In M. Barth, G. Michelsen, M. Rieckmann, & I. Thomas (Eds.), *Handbook of higher education for sustainable development* (pp. 241–260). London: Routledge.

World Bank. (2007). Restoring China's Loess Plateau. Retrieved from http://www.worldbank.org/en/news/feature/2007/03/15/restoring-chinas-loess-plateau

World Commission on Environment and Development (WCED). (1987). *Our common future*. Oxford: Oxford University Press.

CHAPTER 3

A COMPARATIVE ANALYSIS OF APPROACHES TO INTEGRATING SUSTAINABILITY INTO THE CURRICULUM AT A UNIVERSITY IN A SMALL ISLAND DEVELOPING STATE IN THE CARIBBEAN

Alana Griffith and Winston Moore

ABSTRACT

This chapter compares the approaches used in two different disciplines – economics and sociology – to highlight innovative teaching strategies (like the flipped classroom) that are employed to engage students, different ways of integrating sustainability into the curriculum and linking taught units to the sustainable development goals (SDGs) as well as how students confront these issues. The case study courses are delivered at the University of the West Indies Cave Hill Campus in Barbados and the two courses are "Caribbean Social Problems" and "Economic Planning." This chapter demonstrates the necessity of integrating sustainability and the SDGs into course delivery to ensure that the future development of the Caribbean does not compromise future generations.

Keywords: Sustainability; sustainable development goals; blended learning; sociology; economics; Caribbean; curriculum development; flipped classroom; small island developing states; The University of the West Indies

Integrating Sustainable Development into the Curriculum
Innovations in Higher Education Teaching and Learning, Volume 18, 41–56

ISSN: 2055-3641/doi:10.1108/S2055-364120200000018021

INTRODUCTION

Delineating the "Caribbean" as a region can pose challenges, which often sees the countries comprising the region being grouped with Latin America (Meniketti, 2009). The challenge with this delineation is that the specific needs of Caribbean countries can be overlooked. The Commonwealth Caribbean countries have invested significantly in tertiary education, more specifically the University of the West Indies (UWI), which has three main campuses and one online campus. These countries represent a set of small states in the global south used as tourist destinations and present with different levels of development and vulnerability. The CARIBSAVE (2012) Climate Change Risk Atlas assessed the risks and vulnerabilities for 15 Caribbean countries, 12 of which are served by the UWI, and made predictions for each island that included an increase in average atmospheric temperature, sea-level rise, coastal erosion, reduced average annual rainfall, and potential increase for more intense tropical storms. Apart from being vulnerable to environmental shocks, the small open economies of these countries make them vulnerable to international economic shifts. These small island developing states (SIDS) all share common challenges in relation to small size, insularity, remoteness, and proneness to natural disasters (Barnett & Waters, 2016; Briguglio, 1995).

As a higher education institution (HEI) in the Caribbean, the 70-year old UWIs' Mission is to "advance learning, create knowledge and foster innovation for the positive transformation of the Caribbean and wider world." The 17 English-speaking Caribbean countries that contribute to and are served by the UWI include Anguilla, Antigua & Barbuda, The Bahamas, Barbados, Belize, the British Virgin Islands, Cayman Islands, Dominica, Grenada, Guyana, Jamaica, Montserrat, St Kitts & Nevis, St Lucia, St Vincent and the Grenadines, Trinidad and Tobago and Turks and Caicos. The UWI's direct and indirect role in the development of the Caribbean has been documented. To date, it has produced 23 heads of government in the English-speaking Caribbean, one Nobel laureate, two members of the 2007 Nobel Prize winning Intergovernmental Panel on Climate Change, and many other notable alumni (The University of the West Indies (UWI), n.d.-b). Given the UWI's role in higher education in the Caribbean thus far, it is not impossible to believe that among its present 50,000 students may be future leaders, be they heads of government, Nobel laureates, business leaders, innovators, and key persons required for sustainable development in the Caribbean. Recent 2019 rankings by Times Higher Education ranked the UWI as 37th out of 129 of the best universities in Latin America and the Caribbean, in the top 5% of Universities globally and 4th in the criteria for international outlook (Times Higher Education, 2019).

The site for the delivery of the two case study courses used in this chapter is the UWI Cave Hill Campus located on the SIDS of Barbados. The two courses delivered in the Faculty of Social Sciences: Economic Planning and Caribbean Social Problems. These are both final year courses but Economic Planning is an elective housed in the Department of Economics while Caribbean Social Problems is a required/core course taught from the Department of Government,

Sociology, Social Work and Psychology. The comparative analysis of different methods of integrating sustainability into the curriculum, as captured in different disciplinary courses and the "correspondence of topics" acknowledges the correlation between economic and social development. The selection also reflects disciplines related to two dimensions of sustainable development – economy and society. The chapter contrasts the integration of sustainable development in the Caribbean in the curriculum using different disciplinary approaches. Although it is not an institutional requirement that courses integrate the sustainable development goals (SDGs), we understand the importance of integrating the SDGs into the curriculum. It highlights innovative teaching strategies (like the flipped class method) used to engage students, the inclusion of sustainability into the curriculum and linking taught units to the SDGs as well as how students confront these issues. It therefore demonstrates the cross-cutting nature of challenges to sustainable development in the Caribbean Small States that are vulnerable to climate change, natural disasters, and other environmental challenges and demonstrates the necessity of integrating sustainability and the SDGs into course delivery to ensure the future development of the region. There is also a need for twenty-first century education institutions to embrace the collaboration and interdependence of peoples nationally, regionally, and globally (Noddings, 2013). Consequently, while the focus is on the application of the issues to the peculiar circumstances of the region, students are required to situate themselves and the issues covered in the courses as regional and global citizens.

The remainder of this chapter is structured as follows. The "SDGs and HEIs" section summarizes the SDGs and their link to higher educational institutions. The "Incorporating the SDGs into a Sociology Course" section discusses how the SDGs can be incorporated into a sociology courses, and the "Incorporating the SDGs into an Economics Course" section discusses the incorporation of these goals into an economics course. The final section summarizes the main findings from the study and makes further recommendations for the incorporation of the SDGs into the higher education.

SDGs AND HEIs

The 2030 Agenda for Sustainable Development addresses the balance between the environmental, economic, and social factors as three dimensions of sustainable development through 17 interlinked global goals – the SDGs (UN 2030 Resolution). These SDGs embody the hope of UN Member countries in positively transforming the lives of persons globally without some of the adverse impacts witnessed in the past (UN 2030 Resolution). The achievement of these goals is directly and indirectly linked to the curriculum of HEIs, especially those that operate in a developing country context. Blessinger, Sengupta, and Yamin (2018) acknowledge the responsibility of HEIs in shaping the responsible application of scientific knowledge that is harmonious with the environment and its protection. To this end, HEIs are implored to sign the

Declaration on University Global Engagement and treat lifelong learning as a renewable resource (Blessinger et al., 2018). The UWI has already signed the Declaration.

Among the targets for Goal 4 of the UN's SDGs that pertains to "Quality Education" is Target 4.3 that addresses equality of access to quality tertiary education for males and females (UN SDGs Education). Such a target speaks to the principles of the 1945 Report of the Commission on Higher Education in the Colonies Command Paper (Colonial Office, 1945) that led first to the establishment of the University College of the West Indies now the UWI. The intentions therein were to expand the higher education opportunities available to those living in the colonies granting equal access regardless of gender, race, or religious persuasion to satisfy the professional needs "required for the economic and social development of the Colonies" (Colonial Office, 1945, p. 104). The UWI was also to be pivotal in the training of secondary school teachers (Colonial Office, 1945), which can be linked to Target 4.C of the SDGs. Prior to the establishment of the UWI, the provision of tertiary education in the Commonwealth Caribbean was at much lower levels than those in North America, Western Europe, and even Spanish colonies (Miller, 2000). In its present context, it can be said that the role of the UWI in the Caribbean is inextricably linked to not only SDG4 but also the SDGs in general and provides an ideal HEI where this issue can be examined. The cycle of teaching and learning requires that any efforts at sustainable development be included in the curriculum, especially at HEIs whose main aim is not revenue generation but provision of a public good.

As a global university (with centers around the globe conducting teaching and research) but rooted in the Caribbean, the UWI has acknowledged its role in sustainable development and as an institution has created several support initiatives. For instance, its Institute for Sustainable Development aims at improving the resilience of Caribbean islands and sustainability of development efforts. Likewise, its Centre for Resource Management and Environmental Studies (CERMES) focuses on natural resource management in a tropical island environment and the Sustainable Economic Development Unit. In January 2019, the UWI was selected by the International Association of Universities (IAU) to lead in its efforts to achieve SDG 13 (Climate Action) for a climate smart world (The UWI, 2019). The UWI's Office of Development in collaboration with the United Nations Development Programme (UNDP) launched its SDG Youth Advocacy Campaign in April 2017. There have also been student-led approaches to SDG awareness like those by the UWI Cave Hill Guild of Students called "Pelicans of Action" that involves 36 student-led organizations educating students about SDGs and ways in which they can contribute to sustainable development using various activities (UWI Guild of Students). While these efforts exist at a macro- and micro-levels, the processes that occur at the meso-level with regard to attempts by lecturers to integrate sustainability into their courses is not necessarily captured; neither is it mandatory. The ensuing sections provide examples of two such efforts.

INCORPORATING THE SDGS INTO A SOCIOLOGY COURSE

The course Caribbean Social Problems delivered at the UWI Cave Hill Campus is also offered as a franchise course to the Sir Arthur Lewis Community College in St Lucia. Caribbean Social Problems exposes students to a sociological analysis of major issues facing the Caribbean. It assesses the extent, nature, causes, and consequences of social problems in the Caribbean, their links to broader global issues and the influence of supranational agencies like the United Nations (UN) and World Health Organization in helping define and address social problems within the region and beyond. Consequently, students are better able to evaluate the impact of human behavior on societies and efforts being used to address social problems in the Caribbean that adhere to the principles of sustainable development.

Delivered using a flipped classroom approach, several methods are used to engage students and encourage active learning. The flipped classroom is a form of blended learning that not only provides students with a bridge between traditional classrooms and fully online programs but also provides the online platform that is used to facilitate knowledge acquisition before class and thus dedicate face-to-face (F2F) class time to clarification of issues and higher order taxonomic activities like knowledge application and evaluation (Bishop & Verleger, 2013; O'Flaherty & Phillips, 2015). The knowledge acquisition prior to F2F class is facilitated through online videos and video lectures rather than in-class discussions of readings assigned out of class given the limitations that accompany independent reading as an instruction method (Bishop & Verleger, 2013). The flipped classroom method should not to be equated with simply inserting video lectures into course content. While it is arguable whether the method is "new" (Herreid & Schiller, 2013), it represents an innovative approach to instruction and a way to reach students who may wish to obtain a degree and appreciate flexibility but are averse to pursuing fully online studies (Crews & Butterfield, 2014). The method has been shown to positively impact on overall student performance when compared with F2F instruction (Griffith, 2017; Henrie, Bodily, Manwaring, & Graham, 2015; Newman, Kim, Jung Lee, Brown, & Hutson, 2016). Flipping the classroom allows for learning to occur through discussion and debate as a social experience through the small break out groups and larger tutorial groups as well as online discussions. It also contributes to transferrable skills in its use of technology for education. Acknowledging new frontiers for learning, while not relying on mobile technologies, students are encouraged to use smart phones for in-class activities like accessing the course page for class polls/surveys, readings, podcasts, videos, inter alia. The e-Learning (Moodle) platform is accessed by several students using their mobile phones.

Although several methods exist for implementing a flipped classroom, the approach used was one where readings were posted to the e-Learning platform no less than a week before the class along with any videos. The first hour of F2F lectures was spent in peer-to-peer break out discussions comprising approximately five students that addressed three explanatory or application questions on the scheduled topic. During this period, the discussion groups were visited by the lecturer to help "guide on the side" and clarify any unclear issues. These discussions were followed by a brief lecture to further clear up muddy points and reinforce key take away points. Focusing on breaking down large lectures into small breakout groups encourages small-group learning. Readings for online lectures were similarly posted no less than a week before the lecture. The "voice over PowerPoint" video lectures when posted online were not repeated as F2F topics and online discussions and online quizzes replaced F2F tutorials/seminars. In this format, approximately 65% of the course activity moved online with four online lectures out of ten total lectures, three online discussions, and two online quizzes for the academic year 2018–2019. Students also earned 15% of their total grade or 30% of the coursework grade from online assessment methods.

It has been suggested that involving students in the creation of their curriculum not only motivates them to learn the content but has the latent effect of teaching them their voice matters and makes a contribution vis-à-vis possessing educational value (Ornstein & Hunkins, 2018). Furthermore, Bron and Veugelers (2014) contend that the inclusion of students' views in the design and development of curriculum enhances their understanding of their "right to participate" in decision-making. In an effort to engage students and to have them think critically about social problems confronting the Caribbean region, students were involved in shaping the course syllabus for the semester. Assuming students present to the course tabula rasa, the first lecture is generally dedicated to introducing students to the "definition" of social problems and realist and constructionist perspectives on social problems as well as the distinction of personal troubles from public issues. That is introducing students to the work of C. Wright Mills and helping them realize how personally experienced challenges are linked to problems in wider society, thereby, engaging in reflexive practice to understand the social life. At the end of the first lecture, students are invited to suggest issues that can be considered social problems in the Caribbean. These topics are written down on the whiteboard, typed and entered into an anonymous feedback/survey tool on the e-Learning platform where students get two weeks to vote on the two top social problems they wish to study for the semester. This student input is supplementary to general student evaluations and quinquennial curriculum reviews that involve past and present students as stakeholders in assessing and shaping the curriculum. The most popular selections are then added to the core content around theoretical perspectives on social problems. The lecturer was allowed to choose one "wild card" topic that was of importance for the course. Survey results for the most popular choices of the cohort at the center of this study (academic year 2018–2019) are seen in Table 1.

This prioritization in student selections was also seen in the selection of SDGs for the first online discussion. This will be discussed further. The wild

Table 1. Most Popular Topics Chosen by Students in Academic Year 2018–2019.

Topic	Response Rate (%)
Poverty	40
Unemployment	28.5
Health inequalities and non-communicable diseases	22.9
Pollution and climate change	17.1
HIV	14.2
Child abuse	11.4

card topic chosen was crime. The course syllabus then had the following units in which student selections would be addressed: (i) the study of social problems (introductory); (ii) social problems in a theoretical context; (iii) poverty; (iv) unemployment, underemployment, and labor challenges; (v) changing families and family issues (including child abuse and domestic violence); (vi) society and the environment (including social forces affecting climate change and environmental degradation; the relationship between social problems and environmental problems; (vii) health insecurity [including health inequalities, non-communicable diseases and HIV]; and (viii) crime in the Caribbean. Competing theoretical sociology perspectives on each topic was discussed. When one examines the content covered in Caribbean Social Problems, it is evident that the content aligns with the goals of the SDGs. Not only were issues of sustainability included in lectures but in the coursework assignments, which included a written group paper, two online quizzes, two online discussions, and a presentation.

The course had four broad objectives and each unit of study had its own objectives which were to: (1) identify social problems; (2) analyze social problems in the Caribbean using sociological perspectives; (3) explain the interrelationship of some social problems; and (4) assess the impacts of social problems. Throughout the course, special emphasis was placed on the political, economic, and social structures of the society and how they impact/influence/cause social problems. It was therefore imperative that issues around sustainability be integrated into course content. A key objective was demonstrating the links between SDGs and the course. In Unit 2 of the course, in addition to covering broad theoretical perspectives on social problems (conflict theory, symbolic interactionism, structural functionalism, and feminism), students are introduced to the global dimensions of social problems and the impact of supranational agencies like the UN and World Health Organization in identifying such problems and providing support in their amelioration The inclusion of a global dimension in a Caribbean Social Problems course allows for students to situate themselves as global citizens, acknowledging that the forms, causes, and consequences of social problems are not restricted to national boundaries. Such practices are keys to integrate the knowledge systems. It has been suggested that this kind of enrichment in understanding expands our knowledge and empowers other diverse forms of knowledge (United Nations Educational, Scientific and Cultural Organization (UNESCO), 2015). An online "voice over PowerPoint" lecture on social problems in the global context is delivered along with a graded online discussion. The attendant graded online discussion specifically sought to integrate the SDGs into the curriculum by having students reflect on the SDGs and select one goal from among the UN' SDGs. They had to use their selection to describe how a global approach enhances our understanding of and approaches to addressing social problems and the contribution of globalization to our study and experience of social problems. The second online discussion more specifically focused on the SDG16 (Peace, Justice, and Strong Institutions). It required that after watching the lecture students

> [R]ead SDG 16. Choose one of the targets. Using an applicable theory and supporting empirical evidence, explain why there is a need for these targets. Use evidence to support your argument.

The rubric for the online discussion is presented in Table 2.

Of the 57 posts made in the first online discussion, 13 (22%) addressed Goal No. 1 – No poverty. This was unsurprising given the responses for topics. One such response from a student stated:

Table 2. Rubric for Caribbean Social Problems Online Discussion.

	Forum Discussion Grading Criteria		
	0–2 Points Poor	2.5–3 Points Fair	3.5–4 Points Very Good
Quality of post	Post includes some information requested but not focused on the questions posed Post was made more than two days after deadline	Post reflects most of the information and evaluation required but outside source material not cited Post was made within two days of deadline	Post reflects the information and evaluation required and relevant information (reading/examples) with proper course material citation Post was made within the prescribed timeline
	0 Points Poor	0.5 Point Fair	1 Point Very Good
Quality of reply	Response includes no relevant information and is relevant to the original post but omits aspects of observation, comparison and analysis of feelings. Simply compliments	Response is relevant to original post; demonstrates some understanding of the topic; provides some evidence of critical analysis that includes reference to observations and feelings	Response is relevant to original post and demonstrates understanding of the topic; supports position with reference to observation, similarities, differences and feelings in relation to factual information

> It is by becoming interconnected economically politically and culturally, that we are able to approach poverty efficiently by enacting policies and organizing programs that have proven to work. International interaction among countries can generate insight and explanations to poverty so as to make informed decisions (Smith, 2001). Through this widespread knowledge, we can also begin to understand that not all countries are the same thus the same policies that have shown promise, might fail in other regions. According to Kentor (2001), globalization shouldn't be viewed as good or bad. Even though globalization can lead to an increase in foreign dependence, it can also aid in reducing poverty. For instance, trade-openness can ensure increased opportunity for economic activity. All in all, it is through this global amalgamation that we are able to exchange our ideas and work together to explore and resolve any problem we encounter.

Another student proffered that:

> The most important sustainable goal is "No Poverty" by 2030 (The United Nations, 2018). Globalization helps us to address social problems by showing that they are not always nation bound but they are either due to or can be amended by the global reach. For example, foreign investors who prevent the growth of domestic capital or those who supply low wage jobs in the secondary market while repatriating profits contribute to poverty in the host country (Kentor, 2001). Globalization also facilitates and promotes a dependent relationship on powerful countries and their investments. Nevertheless, with global participation "no poverty" is achievable. Due to the speed of migration and capital movement, international work is possible, and poverty can be alleviated. Therefore, as Kentor suggests, globalization is neither good nor bad (Kentor, 2001). The global approach helps us to examine social problems on a wider scale rather than deeming them as local issues and as requiring local solutions.

The emphasis therefore is on finding safe appropriate solutions to problems that affect Caribbean populations.

In discussing SDG2 (Zero Hunger) one student stated:

> This goal sanctioned by the U.N is concerned with ending hunger globally as a part of the 2030 Agenda. A global approach builds awareness to hunger as social problem all over the world and what persons can do to not only help others in need; but also, to develop innovative and economical way to create a safe and consistent food supply without wastage and harm to the environment. Taking a global approach contributes by examining what we can start and stop doing to fix this social problem. Suggestions like growing food, eating more vegetables and less meat, reducing population growth and reducing greenhouse gasses to positively affect climate change are all results of how studies and observations can move toward finding solutions.

In response another student posted:

> Yes I do agree with this discussion as it informs the population of what can be done to fix this social problem and create a safe environment and food for all races of people.

Posts like these address a core tenet of sustainable development which is to think global, act local.

In relation to SDG16 (Peace, Justice, and Strong Institutions), students more directly linked sociological theories with the target and its role in sustainable development. So that one student wrote:

> [...] The point that I will discuss under the SDG16 is, the rule of law and development have significant interrelation and are mutually reinforcing, making it essential for sustainable development at the national and international level. Conflict theorist propose that laws and norms reflect interest of powerful members in society. Karl Marx states that the rule of law plays a big role in the development of western democratic order. Marx believed that the bourgeoisie class maintains discrimination throughout the society as a vehicle to uphold division of class,

> therefore allowing the upper class certain power and relevance. The strain of laws and "norms" placed upon the lower class to attain certain things that the upper class holds has led to crime, due to the fact that crime is an element in attaining these things, which leads to a clear imbalance in society and a continual power for the upper class. Currently in the Barbados there is a high rate of unemployment especially for those of the lower class, this will eventually result in a rise in crime levels as those of the lower class find illegal means and ways to attain what they need. In order to rectify such issue there should be equal laws, and equal social strata. This would lend to a successful, sustainable development of nations.

The discussion posts captured succinctly the way in which students were able to link relevant course content with sustainability and the SDGs.

Likewise, it is captured in another student position that:

> SDG 16, hopes to promote peaceful and inclusive societies for sustainable development, through reducing all forms of violence and related deaths everywhere. Murder rates in the Caribbean are higher than in any other region of the world, and assault rates are significantly above the world average (The World Bank Latin America and Caribbean Region, 2007). Crime is a major concern throughout the world and more importantly, the Eastern Caribbean. A crime is any deviant act that breaks the law of the land (Mustapha, 2009). Through multiple channels, crime and violence threaten the welfare of Caribbean citizens. Beyond the direct effect on victims, crime and violence inflict widespread costs, generating a climate of fear for all citizens and diminishing economic growth. Crime and violence present one of the paramount challenges to development in the Caribbean (A Joint Report by the United Nations Office on Drugs and Crime, 2007). In Barbados, gun violence has risen to significant numbers, leading to the incarcerated juveniles and young adults. A study conducted by Chris Nuttal, in Glendairy, 1995, shows that these correlates to Edwin Sutherland's Differential Association theory. Which states that criminality, is learnt behaviour through interaction with other persons, specifically within "intimate personal groups." Many young persons are socialized and influenced by the values of attainment of wealth and property and learn many ways criminal techniques from peers and family, to do so. Understanding why crime is committed, helps to strengthen relevant national institutions, including through international cooperation, for building capacity at all levels, in particular in developing countries, to prevent violence.

Taking a humanist approach to curriculum allows for the learners to engage in self-reflection (Ornstein & Hunkins, 2018). Through reflective coursework, students were able to consider the importance of sustainability and the SDGs to Caribbean futures. As posited by Ornstein and Hunkins (2018), while on the surface some course content may be limited in the present lives of students, a global and digital world, the future utility of its relevance in another context could be beneficial. Similarly, the Caribbean countries are not new to providing services for foreign jurisdictions. Through its offshore sector, graduates of the UWI exposed to the global context can transfer skills and knowledge to contribute to sustainable development in another context. The following section explores the approach taken in an advanced level undergraduate economics course.

INCORPORATING THE SDGS INTO AN ECONOMICS COURSE

Many of the students who undertake an economics degree at the UWI will pursue careers that intersect with the SDGs. This could be through either working in a developing country or working in a field that intersects with the SDGs, such as decent work and economic growth and climate action to name a few. An economics student, although not recognizing it, is exposed to some of the objectives of the SDGs. In intermediate macroeconomics courses, economic growth and models of growth are the key part of the curriculum while in microeconomics the notions of consumption and production are modeled, albeit at a highly idealized level. In the final year of their program, students would also normally take a course in development economics or be exposed to similar concepts in advanced theory courses.

It is important to note, however, that students pursuing economics degrees are not in any way a homogenous group. Each cohort of economics students is between 100 and 200 students, with differing backgrounds in relation to their home country and outlook in relation to economics. Outside of the core courses in economics (microeconomics, macroeconomics, research methods, and econometrics) students can pursue specializations in finance, quantitative economics, and development economics. Within these areas of specialization, the courses pursued may differ as well as the student's exposure to various concepts related to development. For example, a student pursuing a specialization in development economics is exposed to courses in development economics as well as economic planning while students taking a finance track are not required to take these courses.

The cultural environment of pursuing a degree in economics at a Caribbean-based university must also be considered. Students pursuing degrees at the UWI are largely from SIDS and being aware of the unique circumstances their countries face are influenced to pursue an economics degree. Their backgrounds also influence their expectations as well as the application of economic concepts. This section takes into account these characteristics and discusses how the sustainable development concepts have been incorporated into a course in economic planning.

The purpose of a course in economic planning is to provide students with the skills necessary to develop plans at the organizational, region, or broader macroeconomic level. A fundamental principle for this course is that markets do not always lead to an equilibrium that is desired by citizens. As a result, policymakers can use policies to obtain some superior outcome. In doing so, the policymaker has to ensure that the policies pursued consider the resources available and do not have a negative impact on the environment. It provides practical guidance on how to approach the development of planning documents as well as the style of presentation that best communicates the ideas provided in these documents. By the end of the course, students are expected to be able to formulate economic plans that add economic value to the firm, region, or economy-wide level, distinguish between the tools that can be used to track the performance of the implementation plan, develop capacity building initiatives and identify some of the emerging issues that may impact economic planning.

The course is taught with the use of F2F lectures and supplemented with online lectures, online discussions, case-based small-group discussions, and tutorial discussion. The course is divided into 12 units and each unit is the equivalent of one week's work. The topics covered include strategy, tools for measuring and evaluating performance, analyzing futures, industry clusters and cluster analysis, multisector analysis, capacity building and leadership, spatial planning and facility location planning as well as green strategy. The majority of topics therefore address many of the SDGs, and students are introduced to the concept of the SDGs throughout the course.

In the unit "green strategy," as an example, students are exposed to a number of topics that overlap with various SDGs, including Goal 7 (Affordable and Clean Energy), Goal 8 (Decent Work and Economic Growth), Goal 11 (Sustainable Cities and Communities), Goal 12 (Responsible Consumption and Production), and Goal 13 (Climate Action). Students are exposed to various sustainability models such as developed by Walmart, GE and Amazon. These case studies are also supplemented with various tools such as "Zero Waste," "Energy Management," "Eco-Labels," "Life Cycle Assessment," and "LEED."

In addition to lecture and tutorial assignments on the topic of "green growth," students are also asked to write a report on the issue of green businesses. The question they have to address brings together various aspects of the course such as strategy, greening, and cost-benefit analysis. An illustration of the question they could be required to answer is as follows:

Select a company in your home country and conduct an assessment of strategically positioning the company as a green business. What are the steps that would need to be undertaken to achieve this objective? Would the potential benefits outweigh the costs? Are there any other companies in the industry (either at home or abroad) that have adopted a similar strategy that you can use as a guide?

The question allows the student to begin to confront the topic of what a green business entails and also how greening can be used as element of the firm's strategic position.

The rubric for assessment (Table 3), distributed to students with the project, encourages students to think about how they present the material as well use data as a means of supporting the analysis presented.

The case study by construction encourages students to think about the issues of sustainability and its relevance to businesses. Students normally consider such issues as renewable energy, energy efficiency, waste, and water consumption as part of the assignment. The assignment also encourages students to think about how a green growth strategy can be applied widely across the economy.

DISCUSSION

Although the case studies in this chapter directly assessed the integration of sustainable development and the SDGs using courses that address the economic and social dimensions of sustainable development, it was clear to see that the third dimension of the environment can be integrated and addressed as interdependent and interrelated to the other dimensions, much like the SDGs. HEIs are critical to the achievement of the UN 2030 Sustainable Development Agenda. This is

Table 3. Rubric for Assessment of Case Study Report.

	4	3	2	1
Organization	Information is very organized with well-constructed paragraphs and subheadings	Information is organized with well-constructed paragraphs	Information is organized, but paragraphs are not well-constructed	The information appears to disorganized
Presentation of data to support argument	Data presented are clearly relate to the main topic and support the argument provided	Data presented are relate to the main topic and the somewhat support the argument provided	Data presented are relate to the main topic but does not support the argument provided	Some data are somewhat related to the main topic but does not support the argument provided
Analysis of data utilized in report	The analysis includes several supporting details and supports the argument provided	The analysis includes 1–2 supporting details and/ or examples and somewhat support the argument provided	The analysis includes 1–2 supporting details and/ or examples but does not support the argument provided	Little or no analysis is provided and is unrelated to the topic
Sources	All sources are accurately documented in the desired format	All sources are accurately documented but a few are not in the desired format	All sources are accurately documented, but many are not in the desired format	Some sources are not accurately documented
Mechanics	No grammatical spelling or punctuation errors	Almost no grammatical, spelling, or punctuation errors	A few grammatical, spelling, or punctuation errors	Many grammatical, spelling, or punctuation errors

especially so of HEIs like the UWI whose expressed mandate and role is pivotal to the advancement of a group of developing countries like those in the Anglophone Caribbean serviced by the UWI. These case studies demonstrate the ability of HEIs to increase the understanding of students about the most pressing social, economic, and environmental challenges being faced in the world today as outlined in the Declaration on University Global Engagement.

The infusion of sustainability into the curriculum of an undergraduate sociology course was made evident in student discussions. Furthermore, although not prompted student's examination essays also reflected on the SDGs, their achievement and the need to understand and address the various social problems to ensure future development of the Caribbean. This "self-application" was achieved through the structuring of course content and linking topics to issues of sustainability, the SDGs and structuring coursework to assess student's analysis of the topics. On the other hand, the chapter demonstrated the integration of SDGs in the economics course was achieved through the application of the concepts to a green business in the student's home country. In this way, the SDGs were expanded beyond the narrow confines of development economics to broader areas of economics that speak to the issues related to strategy, business, and feasibility. This approach to the SDGs makes the goals less esoteric and they are instead seen as tools that can be used for planning purposes as well as enhancing economic outcomes.

CONCLUSION

It is to always be borne in mind that the past, present, and future cohorts of students at the UWI will potentially be occupying positions of leadership within governments, the public service, businesses and non-governmental organizations both domestically and internationally. These roles require an appreciation of the deleterious impact of economic, social, and environmental problems on the future of Caribbean societies. The SDGs are important in a University context and universities are also critical to their achievement. Learning about other contexts contributes to global sustainability. Sharing the UWI perspective is important as it captures the ways sustainability can be integrated into the curriculum at a macro-level through dedicated institutes, centers and units; at a meso-level through individual courses and the efforts like those in this chapter; and at the micro level through student-led education efforts. The UWI could consider offering a Massive Open Online Courses (MOOC) aimed at educating about SDGs for SIDS, coordinated by the UWI and perhaps delivered through the SDG academy, which is an online platform dedicated to online education about sustainability and the SDGs. Mobile technologies can be used to reach other territories, especially those underserved communities either through fully online or blended modalities. The SDGs allow not only for a global development agenda but research such as this facilitates the repurposing of education, especially in HEIs.

ACKNOWLEDGEMENT

The authors wish to thank the students of SOCI3035 for allowing the use of their online discussions as examples in this chapter.

REFERENCES

Barnett, J., & Waters, E. (2016). Rethinking the vulnerability of small island states: Climate change and development in the Pacific Islands. In J. Grugel & D. Hammett, *The London Palgrave handbook of international development* (pp. 731–748). London: Palgrave MacMillan.

Bishop, J., & Verleger, M. A. (2013, June 23–26). The flipped classroom: A survey of the research. Paper presented at the 120th American Society for Engineering Education Annual Conference & Exposition, Atlanta, GA.

Blessinger, P., Sengupta, E., & Yamin, T. S. (2018, November 30). Higher education for a sustainable future. *University World News*. Retrieved from https://www.universityworldnews.com/post.php?story=20181128082326995. Accessed on February 1, 2019.

Briguglio, L. (1995). Small island developing states and their economic vulnerabilities. *World Development*, *23*(9), 1615–1632.

Bron, J., & Veugelers, W. (2014). Why we need to involve our students in curriculum design: Five arguments for student voice. In D. J. Flinders & P. Bruce Uhrmacher, *Curriculum and teaching dialogue* (pp. 125–139). Charlotte, NC: Information Age Publishing.

CARIBSAVE. (2012). 2009–2011 The CARIBSAVE Climate Change Risk Atlas (CCCRA). Retrieved from https://www.caribbeanclimate.bz/2009-2011-the-caribsave-climate-change-risk-atlas-cccra/. Accessed on February 25, 2019.

Colonial Office. (1945, June). Report of the commission on higher education in the colonies. Command Paper 6647.His Majesty's Stationery Office, London. Retrieved from https://www.bcn.cl/obtienearchivo?id=documentos/10221.1/29331/2/213787.pdf. Accessed on January 10, 2019.

Crews, T., & Butterfield, J. (2014). Data for flipped classroom design: Using student feedback to identify the best components from online and face-to-face classes. *Higher Education Studies*, *4*(3), 38–47.

Griffith, A. (2017). The feasibility of flipping: An exploratory analysis of the flipped class in Caribbean sociology. *Journal of Effective Teaching*, *17*(2), 72–89.

Henrie, C. R., Bodily, R., Manwaring, K. C., & Graham, C. R. (2015). Exploring intensive longitudinal measures of student engagement in blended learning. *International Review of Research in Open and Distributed Learning*, *16*(3), 131–155. Retrieved from https://www.researchgate.net/publication/279930877_Exploring_Intensive_Longitudinal_Measures_of_Student_Engagement_in_Blended_Learning. Accessed on February 20, 2019.

Herreid, C., & Schiller, N. (2013). Case studies and the flipped classroom. *Journal of College Science Teaching*, *42*, 62–66.

Kentor, J. (2001). The long term effects globalization. *Journal of Social Problems*, *48*(4), 435–455.

Meniketti, M. (2009). Boundaries, borders, and reference points: The Caribbean defined as geographic region and social reality. *International Journal of Historical Archaeology*, *13*(1), 45–62.

Miller, E. (2000). Access to tertiary education in the Commonwealth Caribbean in the 1990s. In G. Howe (Ed.), *Higher education in the Caribbean: Past, present and future directions* (pp. 117–141). Kingston, Jamaica: The University of the West Indies Press.

Mustapha, N. (2009). *Sociology for Caribbean students*. Kingston, Jamaica: Ian Randle Publishers.

Newman, G., Kim, J., Jung Lee, R., Brown, B. A., & Huston, S. (2016). The perceived effects of flipped teaching on knowledge acquisition. *The Journal of Effective Teaching*, *16*(1), 52–71.

Noddings, N. (2013). *Education and democracy in the 21st century*. New York, NY: Teachers' College Press.

O'Flaherty, J., & Phillips, C. (2015). The use of flipped classrooms in higher education: A scoping review. *Internet and Higher Education*, *25*, 85–95.

Ornstein, A. C., & Hunkins, F. R. (2018). *Curriculum: Foundations, principles and issues* (7th ed., Global Edition). Harlow: Pearson Education Limited.

Smith, D. (2001). Globalization and social problems. *Journal of Social Problems*, *48*(4), 429–434.

The University of the West Indies (UWI). (n.d.-a). About us. Retrieved from https://www.cavehill.uwi.edu/About/. Accessed on February 24, 2019.

The University of the West Indies (UWI). (n.d.-b). Heads of Government. Retrieved from https://www.uwi.edu/alumnionline/points-pride/heads-government. Accessed on June 5, 2019.

The University of the West Indies (UWI). (2019). The UWI selected to lead climate-smart world. Retrieved from https://sta.uwi.edu/news/releases/release.asp?id=21881. Accessed on June 14, 2019.

Times Higher Education. (2019). World university rankings. The University of the West Indies. Retrieved from https://www.timeshighereducation.com/world-university-rankings/university-west-indies. Accessed on January 12, 2019.

United Nations. (n.d.). About the sustainable development goals. Retrieved from https://www.un.org/sustainabledevelopment/sustainable-development-goals/. Accessed on January 21, 2019.

United Nations. (n.d.). Sustainable development goals: Quality education. Retrieved from https://www.un.org/sustainabledevelopment/education/. Accessed on February 5, 2019.

United Nations Educational, Scientific and Cultural Organization (UNESCO). (2015). *Rethinking education: Towards a global common good?* Paris, France: UNESCO. Retrieved from http://www.unesco.org/new/fileadmin/MULTIMEDIA/FIELD/Cairo/images/RethinkingEducation.pdf. Accessed on March 1, 2019.

United Nations General Assembly. (2015). Transforming our world: The 2030 Agenda for Sustainable Development. Resolution adopted by the General Assembly on 25th September 2015. Retrieved from http://www.un.org/ga/search/view_doc.asp?symbol=A/RES/70/1&Lang=E. Accessed on December 11, 2018.

United Nations Office on Drugs and Crime and the Latin America and the Caribbean Region of the World Bank. (2007). *Crime, violence, and development: Trends, costs, and policy options in the Caribbean*. Report No. 37820. Retrieved from https://www.unodc.org/pdf/research/Cr_and_Vio_Car_E.pdf. Accessed on June 25, 2019.

UWI Guild of Students. (n.d.). Pelicans of action: A project by the guild of students UWI Cave Hill. Retrieved from https://sdgactionawards.org/initiative/2420. Accessed on February 20, 2019.

CHAPTER 4

INTEGRATING SUSTAINABLE DEVELOPMENT INTO THE CURRICULUM: A CASE STUDY ON THE DEVELOPING OF SUSTAINABILITY COMPETENCIES IN INDUSTRIAL DESIGN STUDENTS AT A BACHELOR LEVEL IN MEXICO

Martha Elena Núñez López, Robert Huddleston and Roberto Pablo Martínez Lozano

ABSTRACT

This chapter presents a case study on integrating sustainable development (SD) into the Industrial Design Bachelor's course at Tecnologico de Monterrey (TEC) in Mexico. The research is being conducted at TEC, where the lead author is a Professor of Architecture. Mexico has a five-year national development plan: the "Plan Nacional de Desarrollo 2013–2018" (PND). This provides a basis for guiding the policies and programs of the Government of Mexico. The focus of this study is on the PND's "quality education goal to make scientific, technological, and innovation development pillars for

Integrating Sustainable Development into the Curriculum
Innovations in Higher Education Teaching and Learning, Volume 18, 57–71

ISSN: 2055-3641/doi:10.1108/S2055-364120200000018022

sustainable economic and social progress." This case study investigates a curriculum intervention, utilizing interviews with students to gather and analyze their responses to the university's development of sustainability competencies. Their responses are explored through comparing a traditional semester with a semester in which sustainability contents and assessment criteria were added to the curriculum of the Industrial Design Workshop courses. The results reveal that the students recognized a significant advance in their development of sustainability competencies and that this had resulted from this curriculum intervention. This chapter proposes that the findings of the study indicate that a holistic approach has the potential to contribute significantly to SD education in Mexico.

Keywords: Sustainability; sustainable development education; competencies; industrial design; higher education; curriculum

INTRODUCTION

The lead author's current study addresses the integration of sustainable development (SD) into the Industrial Design curriculum at Tecnologico de Monterrey (TEC). SD education (SDE) requires the integration of sustainability content, such as the consumption of natural resources, into the curriculum. It also necessitates changes in learning, teaching, and assessment methods and participatory approaches, which can engage and motivate students and result in actions toward a sustainable future. It is worth noting here that United Nations Educational, Scientific and Cultural Organization (UNESCO) (2005) was the organization responsible for coordinating the 2005–2014 Education for Sustainable Development (ESD) of the United Nations.

The study, which forms the basis of this chapter, contributes to the university's Strategic Plan 2020. This plan involves the renewal of curricula together with the development of learning, teaching, and assessment methods. In this context, the study specifically tests and evaluates students' and professors' responses to the creation and application of sustainability competencies. Competency-based education (CBE) and its relevance to SDE at TEC is discussed later in this chapter in the relation to the method of the study ("Situating the Case" of the "Method" Section).

On the topic of sustainability, the International Higher Education Teaching & Learning Association (2018) challenged universities to:

> keep up with the ever-changing world and develop students' ability to appreciate the complexities of sustainability and translate their knowledge into systemic, anticipatory, and critical thinking ability.

The study's research methods are designed in order to investigate the interrelationships between the participants' different environments and experiences, for example, how the campus is integral to the curriculum. This environmental dimension is explored by using "walking interviews," a method employed in social sciences, geography, and ethnography (Clark & Emmel, 2010). The method selected for this aspect of the study transcribes and photographically documents the participants' visual and oral responses to their experiences of their campus environment in relation to SDE. The findings indicate that this method provides opportunities for participants to communicate their perceptions of the curriculum more holistically (Clark & Emmel, 2010). An expert in this methodology, Penelope Kinney (2017, p. 2), of Otago Polytechnic, describes how a "toolkit" developed by Emmel and Clark (2009, pp. 1–24) offers researchers who are conducting walking interviews "insights into what [they] should consider." Kinney (2017, p. 2) explains that an important advantage is that participants can take decisions, "they are in control of the interview" and "regarded as experts in their geographical area." In this way, the study's use of "participatory walking interviews" (Kinney, 2017, p. 2) provides a range of responses to the curriculum, as conceived above: as the sum of the students' learning experiences. These responses complement the data gathered through interviews and questionnaires in ways that the authors believe are necessary to address the need for holistic change in SDE, defined in the context of this study as bringing together, through the design of the new curricula at TEC, education practices, environments, behaviors and the mind-set of learners and educators in the field of SD. Through initiating, testing and evaluating curriculum interventions, this study seeks to answer the call for responsible actions, based on rational principles and ethical values, and for embedding SD within the curriculum (Strang, 2010).

According to the UNESCO (2010), "Sustainable development concerns a process of change," which "evolves as we learn to grasp its wide implications for all aspects of our lives." With respect to this process of change, SDE must equally be able to adapt and evolve and it is imperative that the curriculum and a range of interconnected education approaches are equipped to engage in and contribute constructively to such a process. In this regard, this study's location in Mexico addresses the relationship of SDE to local contexts, as they emerge and evolve.

Today, there is general recognition of the importance of conserving natural capital and environmental goods and services, as conservation is a key to the development of countries and the welfare of the population. In 1994, environmental concerns led the Environmental Protection Agency (EPA) in the United States to begin making recommendations on how schools and universities could teach SD. The Center for Green Schools (1994) made recommendations on the teaching of SD, with the aim of enabling all students to graduate with a better understanding of environmental issues. Through exploring ways of innovating learning, teaching and assessment, this research addresses the call for education to play a significant role in improving and continuously supporting sustainability education.

Ananda, Bisaillonb, Websterb, and Amor (2015, p. 917) propose working on sustainability issues at a regional level, as this affords students and educators "exposure to practically dealing with sustainability issues and challenges in the society" and thereby "to be able to see sustainable solutions being adopted." For Ananda et al. (2015, p. 917), working at a regional level also enables a group to be active "beyond the educational institute" and an opportunity to meet, network, and collaborate with people with a common vision and thereby to benefit from peer review of the integration approach. It is important to consider this perspective (or point of view) in the context of Mexico. Mexico has "the lowest number of higher education students among the OECD countries" (Mexicanos primero, 2016, p. 1). As stated in the 2016 "Mexican First" report ("Reprueba Mexico en posgrados"), only 17% of Mexicans study at a higher education level, 4% progress to graduate studies, and less than 1% go on to study at PhD level. Given these figures, the capability of Mexico's universities to positively affect the country's SD may be questioned. However, the authors argue that Mexican universities can and must be active in supporting the country's SD by acting "beyond the educational institute," a concept that is also proposed by Sevra Davis, Head of Learning, The Design Museum, London: "It is important to make Universities and business work together" (personal communication with the lead author, March 1, 2019). That is, for SDE to contribute significantly to achieving Mexico's "quality education goal (PND)," universities must work together with businesses and professional organizations, for example, through companies and employers participating in SDE through the design and development of SD competencies.

This study has three levels of engagement and contribution: (1) practical curriculum changes at TEC; (2) education strategy at TEC, aligned to the PND; and (3) holistic change to support SDE.

Considering SD as a new way to see the world and take action, the authors' review of literature for this study indicates that the integration of SD into higher education requires a range of fundamental and practical changes (Mexicanos primero, 2016). Some examples of changes that scholars have proposed are the following: Stebbing and Tischner (2015, p. 118) propose that a "behavior change is a big question"; that "status and conspicuous consumption are unsustainable" and that the "old design paradigm" has contributed to "unsustainable attitudes," "wasteful practices," and a "throw-away-society" (Stebbing & Tischner, 2015, p. 24). Otto (2007, p. 3) analyses the notion of cultural change, proposing three pillars of sustainable design: social, environmental, and economic. Rogel, Rodriguez, and Macías (2015, p. 9) proposes a fourth one: the political, and Edwards (2005, p. 2), expands upon this proposal by identifying a connection between the environmental impact of products during their life cycle and the social aspects of manufacture.

Murray (2011, p. 1) proposes, with respect to SD, that it is important to "personalise sustainability"; that people need to understand sustainability but they also to need to "connect with it at a personal level," in order to generate a sustainable change. This emphasis on the individual means that sustainability education needs to inform students so as to enable them to modify their attitudes, behaviors, and actions with regard to environmental responsibility. In this respect, the earlier students are exposed to sustainability education, the easier it will be to instill environmentally friendly habits and opinions (Advanced Materials & Processes, 2015).

Watkins (2013, p. 133) proposes the application of workshops, the "Rethinking Design series," in order to introduce students to the "wider social aspects of Sustainable Product Design," through the use of audio visual group-based workshops. He states that the workshops enable a "learning environment where a deep understanding of the social aspects of Sustainable Product Design could be developed" (Watkins, 2013, p. 20).

According to Tromp, Hekkert, and Verbeek (2011, p. 3), the conventional goal of sustainable design was designing products that could be recycled and that could use less energy to be produced and used. However, since then and according to more recent angles of approach and goals "to really effect change, sustainable design must be capable of changing user behavior." Industrial designers who develop the skills necessary to create products with a sustainability culture during their studies could directly or indirectly influence their users.

The above perspectives reflect the multifaceted and complex nature of SD and ESD, and they exemplify the existence of a range of concepts and approaches. In this context, the emphasis of this study is not on prescribing the contents of SD curricula, but centers on developing, applying, and evaluating pedagogical instruments and approaches, at a strategic level, and in ways that can best enable holistic change and the integration of SD at TEC.

AIM

The authors argue that scholars in the discipline of Industrial Design (alternatively named Product Design) have not yet adequately addressed the creation of effective learning approaches in relation to the challenges of the PND and its progress measurement indicators. This investigation aims to answer the question: What actions are required to support a product design curriculum, capable of addressing the challenges of sustainable economic and social progress in Mexico?

The study's primary aims are: (i) to integrate SD into the curriculum; (ii) to evaluate and measure the effectiveness of this development; and (iii) in order to participate actively in SD's "process of change," to embed SDE into TEC's instruments for continuous curriculum development. As mentioned above, the study also recognizes the need for a mind-set change among ESD stakeholders. Although this ambition may be beyond the timescale scope of the study, the authors believe that the study's aims, as discussed above, can provide the means to work progressively toward positively affecting changes in attitudes, behaviors, and ways of thinking with respect to SD in Mexico.

METHOD

Situating the Case

With the aim of improving SDE in Mexico, the study applies empirical research methods in relation to the Product Design Bachelor's at Tecnologico de Monterrey, Mexico. As a cornerstone of Tecnologico de Monterrey's Strategic Plan 2020, in 2014, the university set out its Tec21 "Educational Model" and the process of renewing the curriculum, following in 2017 by the conformation of a team of professors from different Tec campuses for the detailed design of the 2019 study plan, which will begin to be offered as of August 2019. The main difference of this new curriculum in comparison with those previously offered in Tecnologico de Monterrey, lies in the methods of learning and evaluation based on competencies.

The moment of this development is superimposed with the SDE study of the lead author that began in 2016 and is planned to be completed in 2020. This time scale allows to validate the result of the research and makes adjustments during the time of implementation of the new curriculum.

The authors propose that CBE has the potential to provide an important pedagogical strategy and method in the context of the "Tec21 Educational Model" and the SDE aims of this study. The rationale behind this thinking is that CBE enables greater emphasis on the individual student's personal learning journey, including their motivations and attitudes with regard to real-world design problems and contexts. Also, through identifying a range of competencies, it is considered that this method can better encourage, within a student, their ability to engage critically and responsibly with processes of change, as a professional and/or in their personal lives.

According to Guzmán (2003, p. 159), an advantage of the application of a CBE model is that it prevents the contents of curricula from "responding to ... disciplinary whims or is the result of pressure from the institution's power groups." A whim, in this context, could arguably be thought of as a potentially constructive starting point for innovation. However, the key point with respect to the current study's focus on integrating SDE in to the "Tec21 Educational Model" is that, as Guzmán goes on to state, competencies enable "the development of a set knowledge, skills and attitudes" that "allows the learner to acquire integral and relevant skills for their professional work (Guzmán, 2003, p. 145)."

Once an educational model has been developed and implemented it is important that SDE actors need to know what is happening, and one way of doing this is by measuring the effectiveness of SDE within their courses and learning environments as a whole. As identified above, one of the aims of the study is to analyze and evaluate the student SD learning by developing and testing the application of the competencies, aligned with the Tec21 Educational Model.

As discussed above, the development of competencies is a central instrument of the Tec21 Educational Model. Here, a brief summary of the education contexts that have led to this development is relevant to understanding the relationship between competencies, the university's education goals and within this scope, the integration of SD. In the Tecnológico de Monterrey's digital magazine, *EduTrends*, periodic reports are presented on educational developments that demonstrate the greatest potential for impact in higher education. The *EduTrends* report published in February 2015 was dedicated to CBE and it commences by explaining that the concept of competency in education appeared in the 1970s, in response to the inadequate relationship that existed at that time between education programs and the needs of the labor market EduTrends (2015).

In Mexico, an official policy related to CBE emerged in 1994 through the Project for the Modernization of Technical Education and Training (PMETYC). It focused on vocational training in charge of the institutions of the Technological Education Subsystem, such as Colegio Nacional de Educación Profesional Técnica (CONALEP) and the National Polytechnic Institute. In 1995, the Council for Standardization and Certification of Labor Competence (CONOCER) emerged. Based on a model from England, it included representatives of private and social sectors and members of the public. In 2004, the educational model of the Technological Universities was implemented, in order to respond to the needs of the different productive sectors and society.

Tecnologico de Monterrey has decided to use this education model for the 2019 curricula because of its closer connection to and emphasis on professional competencies and its alignment to the trends of higher education worldwide. In the Educational Model Tec21, "the evaluation focuses on the development of student competencies that make up their graduation profile" and the achievement of these "at key moments of their academic life" is observed (EduTrends, 2015).

The Tec21 Educational Model proposes a change in the curriculum and forms strategy, which has led to redefining the learning assessment process consistently with the competency-based approach (EduTrends, 2015). Among the advantages of CBE, is that this model represents an alternative to the current model, which articulates learning objectives: that is, students no longer focus only on passing the course (know how), but must demonstrate the acquisition of a set of skills, values, and attitudes (know how to be). Another advantage is that CBE is active, functional, and comprehensive; and brings the students to real challenges where they must act in ways that enable the practical application of the knowledge they have acquired. The advantage of TEC's development of competency-based learning, teaching, and assessment is that it makes it possible to include knowledge, different skills and attitudes, for example, cognitive skills, such as critical and problem-solving skills and also that by encouraging the students to construct knowledge, they can be more active in their learning.

Curriculum Intervention and Its Methodology

The 2017 Industrial Design Bachelor's curriculum at Tecnologico de Monterrey comprises nine semesters. The Industrial Design Workshop courses run continuously from the third to the ninth semesters. Through the authors' study, an intervention was applied to these courses. Industrial Design Workshop courses were selected because of their integrative purpose within each semester. For example, these courses are where the students apply, in practical ways, the contents of the rest of their courses, comprising ethnography, user centered design, representation techniques, materials and manufacturing processes, models and prototypes, and digital modeling.

Curriculum Intervention Methodology

It is worth noting that the original permission that the researcher requested for the intervention was for the semester January–May 2019; however, the Director of the Industrial Design Academic Department considered it important to begin the intervention as soon as possible given its relevance to Sustainability and its potential to inform the Design of Study Plans 2019 team, of which he himself is a member. At the end of the semester, the students' perceptions on the development of competencies, linked to SD, were evaluated through conducting a survey. The students were asked to give their opinion on the extent to which they developed the competencies during that semester – with competencies (August–December 2018), and during the previous semester – without competencies (January–May 2018). The survey used a 1–5 scale, where 1 is low and 5 is high.

The Development of Competencies

Five SD competencies were drafted based on the opinion of different expert professors from the Puebla campus, some of them belonging to the team that is leading the 2019 study plan design at Tecnologico de Monterrey. The authors presented the results of this process to the Director of the Industrial Design Academic Department at the Puebla campus, who approved the application of the intervention. As described above, the intervention included the participation of the professors who were teaching the subjects of Design Workshops during the August–December 2018 semester.

The five declared competencies for this study are the following:

1. The student's work demonstrates the knowledge and understanding of SD issues locally/globally.
2. The student's work demonstrates a mind-set change in relation to SD through the successful application of the principles to her/his projects as well as in her/his own lifestyle.
3. The student assesses the life cycle of the product/s of her/his projects based on criteria that relate to the minimization of the multiple environmental impacts throughout the cycle.
4. The student encourages the rest of the group's involvement in SD when presenting her/his projects to the group.
5. The student looks for and finds knowledge inputs and expertise from other disciplines in order to meet the sustainability requirements that her/his project needs.

In addition to the support requested from the aforementioned experts, the tool called "Roadmap for Integrated Sustainability" created by the United Nations Global Compact, a voluntary initiative to implement universal sustainability principles to support UN goals, was also taken into account in drafting the five competencies for this study. This "world's largest corporate sustainability initiative" provides guidance on integrating sustainability-related goals and strategies across organizations through five stages of sustainability and shows the journey that a company must follow (Secretaría de Educación Pública General Coordination of Technological Universities, 2008).

The surveys were applied in the seven Industrial Design Workshop courses. A total of 72 industrial design students participated in the study.

DATA ANALYSIS AND RESULTS

Based on the students' responses to the survey, when comparing the semester in which the sustainability and assessment criteria were included in the curriculum via the intervention (August–September 2018) to the previous semester, without the inclusion of these criteria (January–May 2018), the results show a marked advance, as a result of the intervention, in the students' perceptions of their development of sustainability competencies.

Survey of Students' Responses to Competencies

The results are as follows:

Competency 1: *The student's work demonstrates the knowledge and understanding of SD issues locally/globally*. When comparing the two semesters, the graphs show an increase in the percentage of students selecting the highest competency development level (level 5) as well as a decrease in the percentage of students selecting the lowest level (level 1). In the January–May 2018 semester (without the curriculum intervention), 33.8% of the students selected the highest competency development level (level 5) and 27.7% selected a level 4. In contrast, in the August–December semester (the curriculum intervention), 52.8% selected a level 5 and 30.6% a level of 4. The sum of the levels 4 and 5 comparing the two semesters can be observed in Table A1. In the January–May semester, 10.8% of the students selected the lowest competency development level (level 1). This was, in contrast to the August–December semester (with the curriculum intervention), in which 1.4% selected this lowest level.

Competency 2: The student's work demonstrates a mind-set change in relation to SD through the successful application of the principles to her/his projects as well as in her/his own lifestyle. When comparing the two semesters, the graphs show an increase in the percentage of students selecting the highest level (level 5) of the competency development as well as a noticeable decrease in the lowest level (level 1). In the January–May 2018 semester (without the curriculum intervention), 29.2% of the students selected the highest level of competency development (level 5) and another 29.2% of them selected a level 4. In contrast, for the August–December semester (with the curriculum intervention), 59.7% of the students selected a level 5 and 23.6% a level 4. The sum of the levels 4 and 5 comparing the two semesters can be observed in Table A2. In the January–May semester, 10.8% of the students selected the lowest level

Table A1. Competency #1 Evaluation Score.

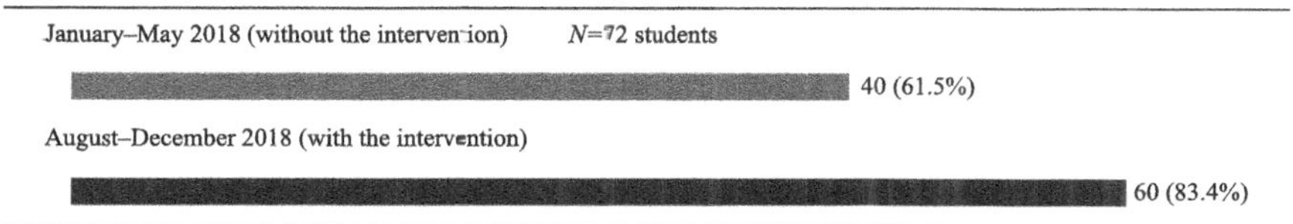

Table A2. Competency #2 Evaluation Score.

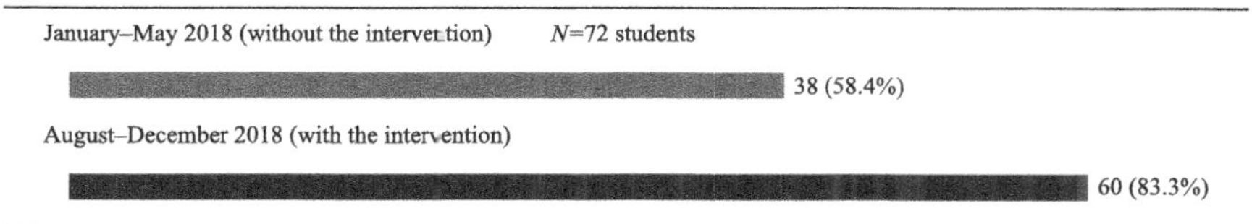

of competency development (level 1). This was in contrast to the August–December semester in which 1.4% selected the lowest competency development level (level 1).

Competency 3: *The student assesses the life cycle of the product/s of her/his projects based on criteria that relate to the minimization of the multiple environmental impacts throughout the cycle*. When comparing the two semesters, the graphs show an increase in the percentage of students selecting the highest competency development level (level 5) as well as a decrease in the lowest level (level 1). In the January–May 2018 semester, 32.2% of the students selected the highest level of competency development (level 5) and 24.6% selected a level 4. In contrast, for the August–December semester (with the curriculum intervention), 52.8% of the students selected a level 5 and 23.6% a level 4. The sum of the levels 4 and 5 comparing the two semesters can be seen in Table A3. In the January–May semester, 10.8% of the students selected the lowest competency development level (level 1). This was in contrast to the August–December semester in which 1.4% of selected the level 1.

Competency 4: *The student encourages the rest of the group's involvement in SD when presenting her/his projects to the group*. When comparing the two semesters, the graphs show an increase in the percentage of students selecting the highest competency development level (level 5) as well as a decrease in the lowest level (level 1). In the January–May 2018, 38.5% of the students selected the highest competency development level (level 5) and 27.7% selected a level 4. In contrast, for the August–December semester (with the curriculum intervention), 56.9% selected a level 5 and 25% selected a level 4. The sum of the levels 4 and 5 comparing the two semesters can be seen in Table A4. In the January–May semester, 10.8% of the students selected the lowest competency development level (level 1). This was in contrast to the August–December semester in which 2.8% selected this lowest level (level 1).

Competency 5: *The student looks for and finds knowledge inputs and expertise from other disciplines in order to meet the sustainability requirements that her/his project needs*. When comparing the two semesters, the graphs show an

Table A3. Competency #3 Evaluation Score.

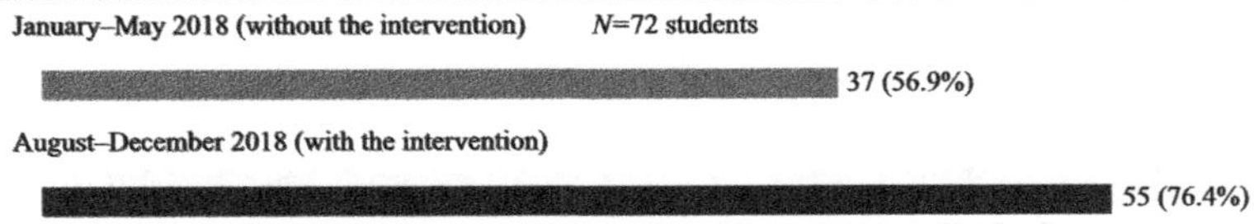

Table A4. Competency #4 Evaluation Score.

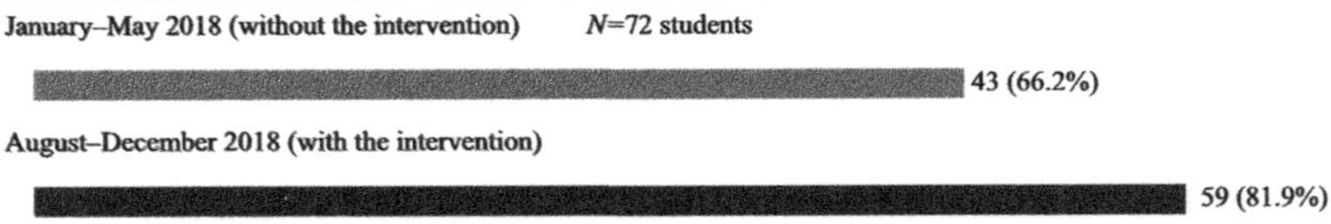

Table A5. Competency #5 Evaluation Score.

January–May 2018 (without the intervention) *N*=72 students

47 (72.3%)

August–December 2018 (with the intervention)

67 (93.1%)

increase in the percentage of students selecting the highest competency development level (level 5) as well as a decrease in the lowest level (level 1). In the January–May 2018 semester, 41.5% of the students selected the highest competency development (level 5) and 30.8% selected a level 4. In contrast, for the August–December semester (with the curriculum intervention), 66.7% of the students selected a level 5 and 26.4% selected a level 4. The sum of the levels 4 and 5 comparing the two semesters can be seen in Table A5. In the January–May semester, 6.2% of the students selected the lowest competency development, level (level 1). This was in contrast to the August–December semester in which 1.4% of them selected the level 1.

Another chart (Table A6) was created in order to summarize the results of the five competencies development with the purpose of enabling visualizing in a

Table A6. Competencies.

	Competency #1	Competency #2	Competency #3	Competency #4	Competency #5
January–May 2018	61.50%	58.40%	56.90%	66.20%	72.30%
August–December 2018	83.40%	83.30%	76.40%	81.90%	93.10%

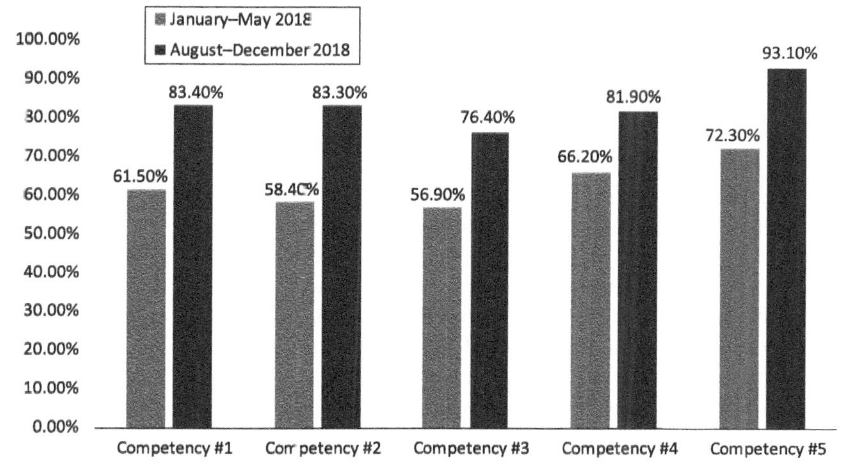

clearer way the contrast between the two semesters. In this chart, the two highest levels (levels 4 and 5) of the competency development were combined in a single value.

DISCUSSION

The Design Workshop courses were considered best suited for testing the extent to which sustainability competencies had been embedded in the curriculum as they offer opportunities for students to apply learning from earlier courses to their design projects. After comparing the students' responses between a traditional semester and a semester in which sustainability contents and assessment criteria were added to the curriculum, the results show that the students recognized a significant advance in their development of sustainability competencies as a result of the curriculum intervention.

The study does not stop here. There is a continuous process of making different iterations and reflecting on the effectiveness of the declared competencies. One reflection that has emerged after collating the results of the first survey application is that the competencies need to be adapted. For example, the competency number #2 "The student's work demonstrates a mind-set change in relation to sustainable development through the successful application of the principles to her/his projects as well as in her/his own lifestyle" is now considered to be a competency that needs to be adapted in order to make it feasible to evaluate incremental changes over the timescale of the study. Revisions to this competency will draw upon aspects of the theoretical material discussed above and "action research" relevant to self-reflective, critical skills (Swann, 2002). A concern with these skills is seen as a way of evaluating the students' engagement with the personal and social responsibilities implicit in design practice and SD. For Swann, action research "demands public accountability and visible self-evaluation."

> There should be no qualms about design benefiting from the need for practitioners to make their processes visible and socially responsible. That's what documented research does, it legitimizes the proposition (Swann, 2002, p. 57).

As referenced earlier, "[s]ustainable development concerns a process of change," which holds "wide implications for all aspects of our lives" (UNESCO, 2010), people need to be able to "personalise sustainability" (Murray, 2011, p. 1) and the use of practical workshops support the development of a "deep understanding of the social aspects of Sustainable Product Design" (Watkins, 2013, p. 20). While the student surveys constitute the study's main method of investigation, the application of "walking interviews," as introduced earlier, enables a broader range of social and political aspects of interest and concern, led by the participants, to inform the study as a whole.

In the next survey application, the students will be asked to accompany each response with comments and supporting evidence. This is considered relevant to ensure that the students are selecting the most accurate value in the survey and that they reflect on the competencies that they have developed. Also, after applying the surveys, a validation process will be undertaken: some of the surveys will be selected randomly to contrast the students' responses with those of the professors' in the development of the competencies through the semester. Discussion sessions are also planned in order to complement the survey results. These will enable the students to consider changes in their thinking and practices as a result of the application of competencies.

A finding that emerged after collating the results of the first survey application relates to the dialogs with the professors who participated in the intervention. In preparation for the intervention, discussions with them explored ways in which they could integrate SD curriculum content and assessment criteria into their classes. In a number of cases, the discussions led to the professors rethinking their perspectives on the relationships of their specialist subjects to SD. In one case, it was observed that for a subject without, at the time, an explicit connection to SD, the professor was open to adjusting the curriculum in order to integrate SD into the curriculum. This finding is indicative of the potential impact of developing a SD-centered curriculum and demonstrates the need for universities, at all levels, to act upon their responsibility to develop, implement, and ultimately embed effective SDE strategies and approaches.

The study will apply and where necessary adjust the same measurement tools, as described above in the section Methodology, in order to continue to gather feedback and measure the effectiveness of the experimental interventions, and from August 2019, the new curriculum has been implemented.

CONCLUSION

The findings of the study so far strongly indicate that the proposed holistic education approach has the potential to contribute, in strategic and practical terms, to improving SDE in Mexico and thereby to achieving the "quality education goals" of Mexico's PND. Through presentations and dialogs with Tecnologico de Monterrey's Industrial Design team, the research is already informing their work on the new curricula to be offered at TEC from August 2019. From this date onward, the work of the Industrial Design team will involve continuing to further develop and refine sustainability competencies and, as described above, to measure the extent to which they progress students' SDE. Finally, rather than stopping with the results obtained in relation to the planned new Industrial Design curricula, the research identifies a series of starting points for further investigations focused on sustainability competencies. It is proposed that future work can be applied to any of TEC's Bachelor's and postgraduate courses, and through the university's national and international academic community and professional connections, broader potential exists for the results to be applied by other Mexican and worldwide universities, according to their local, regional, national, and global contexts.

ACKNOWLEDGMENTS

The authors would like to acknowledge the support of Dr Mary Traynor, Associate Dean (Academic Quality and Student Learning Experience) at Birmingham City University and Dr Geof Hill, Former Professor at Birmingham City University, for their valuable contributions to this study.

REFERENCES

Advanced Materials & Processes. (2015). Materials sustainability app serves as teaching tool. *173*(10), 29–30 Retrieved from https://www.google.com/url?sa=t&rct=j&q=&esrc=s&source=web&cd=1&cad=rja&uact=8&ved=2ahUKEwiAipLIuOflAhVC4qwKHW4lDFEQFjAAegQIABAH&url=https%3A%2F%2Fstatic.asminternational.org%2Famp%2F20151112%2Ffiles%2Fassets2Fcommon%2Fdownloads%2FAMP_DigitalEdition_NovDec.pdf&usg=AOvVaw2q21wshdGe8ZDYPQCHqEyH.

Ananda, C. K., Bisaillonb, V., Websterb, A., & Amor, B. (2015). Integration of sustainable development in higher education: A regional initiative in Quebec (Canada). *Journal of Cleaner Production, 108*, 916–923.

Clark, A., & Emmel, N. (2009). My visual diary. In C. Knowles & P. Sweetman (Eds.), *Picturing the Social Landscape: Visual Methods and the Sociological Imagination* (pp. 1–24). London: Routledge.

Clark, A., & Emmel, N. (2010). 'Using walking interviews'. Realities Toolkit #13, ESRC National Centre for Research Methods, NCRM Working Paper Series. Manchester, UK (pp. 1–24). Accessed on October 20, 2018 Retrieved from https://www.researchgate.net/publication/277803364_Using_Walking_Interviews.

EduTrends. (2015). Educación basada en Competencias (EBC). Observatorio de Innovación Educativa, Tecnologico de Monterrey. Retrieved from https://observatorio.tec.mx/edutrendsradar2015. Accessed on October, 2018.

Edwards, S. (2005). Linking sustainable product design with corporate social responsibility: A consideration of frameworks and measurement tools. *Proceedings of towards sustainable product design*, Farnham, UK (p. 10).

Environmental Protection Agency (EPA). (1994). Environmental Protection Agency Act, 1994 Arrangement of Sections, Part One. Retrieved from http://www.epa.gov.gh/ghanalex/acts/Acts/ENVIRONMENTAL%20PROTECTION%20AGENCY%20ACT%201994.pdf. Accessed on November 3, 2018.

Expansión. (2017). Solo 17 de cada 100 mexicanos ha ido a la Universidad. Retrieved from https://expansion.mx/economia/2017/09/12/solo-17-de-cada-100-mexicanos-han-ido-a-la-universidad. Accessed on October 28, 2018.

Government of the Mexican Republic. (2013). Plan Nacional de Desarrollo 2013–2018, Mexico (pp. 17, 19, 22). Retrieved from http://pnd.gob.mx/wp-ontent/uploads/2013/05/PND

Guzmán, J. (2003, April). Los claroscuros de la educación basada en Competencias. *Nueva Antropología, XIX*(62), 143–162. Retrieved from http://www.redalyc.org/pdf/159/15906208.pdf. Accessed on October 2018.

International Higher Education Teaching & Learning Association. (2018). Emerald sustainable development submission form. Retrieved from https://www.hetl.org/emerald-sustainable-development-submission-form/. Accessed on October 11, 2018.

Kinney, P. (2017). Walking interviews. In N. Gilbert (Ed.), *Social research update* (p. 2). Guildford: University of Surrey.

Mexicanos primero. (2016). Reprueba Mexico en posgrados. Retrieved from http://www.mexicanosprimero.org/index.php/educacion-en-mexico/enterate/noticias-de-hoy/276-reprueba-mexico-en-posgrados. Accessed on October 19, 2018.

Murray, P. (2011). *The sustainable self: A personal approach to sustainability education*. London: Earthscan.

Otto, B. K. (2007). Sustainability. Retrieved from http://www.designcouncil.org.uk/en/About-Design/Business-Essentials/Sustainability. Accessed on October, 2018.

Rogel, E., Rodriguez, S. E. V., & Macías, L. (2015). Educating for sustainable design and competencies. *UACJ*. Retrieved from https://www.researchgate.net/publication/312307472_Educacion_para_un_diseno_sostenible_y_por_competencias. Accessed on October, 2018.

Secretaría de Educación Pública General Coordination of Technological Universities. (2008). United Nations Global Compact 'Roadmap for Integrated Sustainability'. Retrieved from https://www.unglobalcompact.org/take-action/leadership/integrate-sustainability/roadmap. Accessed on October, 2018.

Stebbing, P., & Tischner, U. (2015). Changing paradigms: Designing for a sustainable future. Cumulus Think Tank, International Association of Universities and Colleges of Art, Design and Media. Accessed on October 30, 2018.

Strang, V. (2010). Mapping histories. Cultural landscapes and walkabout methods. In I. Vaccaro, E. Smith, & S. Aswani (Eds.), *Environmental social sciences: Methods and research design* (pp. 132–156). Cambridge: Cambridge University Press.

Swann, C. (2002). Action research and the practice of design. *Design Issues*, *18*(2), 29.

Tecnologico de Monterrey. (2018, October). Modelo de evaluación del aprendizaje (Folleto Morado), pp. 7–14.

Tromp, N., Hekkert, P., & Verbeek, P.-P. (2011). Design for socially responsible behavior: A classification of influence based on intended user experience. *Design Issues*, *27*(3), 3.

U.S. Environmental Protection Agency. (2013). Retrieved from https://www.theguardian.com/environment/epa. Accessed on October 20, 2018.

United Nations Educational, Scientific and Cultural Organization (UNESCO). (2005). Educación para el Desarrollo Sostenible (EDS). Retrieved from http://www.unesco.org/new/es/our-priorities/sustainable-development/. Accessed on October 20, 2018.

United Nations Educational, Scientific and Cultural Organization (UNESCO). (2010). Understanding sustainable development. Retrieved from http://www.unesco.org/education/tlsf/mods/theme_a/mod02.html. Accessed on October 29, 2018.

Watkins, M. (2013). *An investigation into effective methods for teaching social sustainability within product design in British and Irish Universities*. Ph.D. thesis, Loughborough University Institutional Repository.

PART II

INNOVATIVE APPROACHES

CHAPTER 5

PROGRAMS, WORKSHOPS, RESOURCES, AND OTHER SUPPORTS FOR POST-SECONDARY SUSTAINABILITY EDUCATORS

Alice Cassidy, Yona Sipos and Sarah Nyrose

ABSTRACT

There is a growing need to train and support educators to introduce or enhance aspects of sustainability into post-secondary curriculum. The authors provide an overview of integration of curricular sustainability development and education as well as related institutional leadership at the post-secondary level. Turning to educational development for sustainability education, the authors share tools and resources to support educators from any discipline, to introduce, integrate, and/or enhance sustainability in their course, program, or initiative. The authors found very few examples of workshops to post-secondary teachers. For one such example, the Sustainability Education Intensive, a three-day workshop that the authors designed and led at the University of British Columbia. The authors summarize the workshop aspects that two years of participants found helpful, and how workshop involvement affected them as sustainability educators. The authors encourage post-secondary institutions to provide support in the form of workshops, resources, and funding to help educators introduce or enhance aspects of sustainability into their courses and programs. Students are asking for this, and, as they are future leaders, it is important that educators address the numerous environmental, social and economic issues that demand attention.

Integrating Sustainable Development into the Curriculum
Innovations in Higher Education Teaching and Learning, Volume 18, 75–89

ISSN: 2055-3641/doi:10.1108/S2055-364120200000018023

Keywords: Sustainability; sustainable education; educational development; interdisciplinary learning; authentic learning, workshop design

INTRODUCTION

There is a growing need to train and support educators to introduce or enhance aspects of sustainability into the post-secondary curriculum. The World Health Organization (WHO; https://www.who.int/emergencies/ten-threats-to-global-health-in-2019), in launching a five-year strategic plan, underscores the importance of bringing aspects of sustainability into post-secondary education. WHO considers air pollution and climate change to be the greatest environmental risks to human health in 2019.

Brinkhurst, Rose, Maurice, and Ackerman (2011) speak to the importance of faculty and staff as critical leaders in achieving lasting progress toward campus sustainability, an important component in the transition to sustainability more broadly. They also call for greater attention to the potential of faculty and staff and how their work can be supported. In our experience, graduate and undergraduate students also play important roles as sustainability leaders and need support.

Findings of an online survey of over 1,700 students conducted by Beasy, Peterson, Tomlinson, and Tiernan (2016) at the University of Tasmania showed that, while 78% of students recognize that sustainability is important to have in their courses, 69% are unsatisfied with or unaware of sustainability in them. They suggest professional development support to help teachers incorporate sustainability into all courses.

Purposely integrating sustainability can be a tall order unless programs, resources, and other supports are in place. Maloni, Smith, and Napshin (2011) agree offering insight into how to build faculty support for sustainability programs. Filho et al. (2017), surveying 269 experts from 45 countries, found that the greatest barrier to innovation in sustainability practices in universities is lack of support at administrative, committee, and technological levels. O'Flaherty and Liddy (2018), conducting a literature review of education for sustainable development and related topics, note pedagogies and curricular approaches, but not support systems.

This chapter describes programs, workshops, resources, and other supports for faculty, staff, graduate, and undergraduate students. With a focus on educational development, we share tools and resources to support educators to introduce, integrate, and/or enhance sustainability in their course, program, or initiative. For one such example, the Sustainability Education Intensive (SEI), a three-day workshop we designed and led at the University of British Columbia (UBC), we summarize the workshop aspects that two years of participants found helpful, and how workshop involvement affected them as sustainability educators.

There are many definitions of the term sustainability, commonly including social, economic, and environmental aspects. The World Commission on Environment and

Development (1987), also known as the Brundtland Report defined sustainability as "meeting the needs of the present without compromising the ability of future generations to meet their own needs." We provide eight other definitions of sustainability at https://cassidyinview.wordpress.com/teaching-and-learning-workshops/sustainability-education-intensive-a-workshop-for-post-secondary-teachers/ used in a SEI activity.

AN OVERVIEW OF SELECTED LITERATURE

We (Cassidy, Sipos, & Nyrose, 2014) have previously provided examples of sustainability education worldwide, focusing on degree programs and programs on sustainability education for post-secondary educators. We noted much potential for future work, including a hope for more literature describing support for sustainability educators and its efficacy. Here, we provide an overview of selected literature on integration of curricular sustainability development and education and related institutional leadership. We focus on post-secondary education with an emphasis on Canadian examples.

Table 1 summarizes much of the literature we found, with examples at cross-institutional and institutional levels, as well as other supports. Cross-institutional examples include networks, studies and strategic plans, as well as programs and resources that can be used anywhere. Institutional examples include those within a course or program, studies of an institution, and resources available to the wider community. Some examples describe opportunities to take part in research or education. In some cases, programs support faculty as well as students and others in the academic community. Two of us (Cassidy & Sipos, 2007) have emphasized the importance of students as co-researchers and co-presenters.

Though some examples in Table 1 focus on aspects of sustainability within particular disciplines or contexts, we suggest that the principles and strategies can be applied to other areas. For example, O'Connell, Potter, Curthoys, Dyment, and Cuthbertson (2005) encourage outdoor recreation educators to integrate aspects of sustainability into their curricula. This is echoed by Lugg (2007), noting that the pedagogical approach of outdoor learning "lends itself to holistic and experimental learning and enables integration of knowledge and skills from a range of discipline areas" (p. 97). Lefebvre's (2000) criteria for evaluating sustainability education pedagogy include interconnections between social, environmental, and economic aspects, emphasizing interacting and learning with nature. We feel that these approaches are applicable to other pedagogies and contexts.

Additional resources – over 400 and growing – including journals and journal publications, conference sessions, books, programs, and initiatives related to sustainability education and leadership are found at https://cassidyinview.wordpress.com/by-topic/sustainability/sustainability-education-and-leadership/. Funded by the Educational Developers Caucus, this repository was created with the help of a graduate student and an undergraduate student. A searchable site includes 800 resources related to learning and teaching alone https://www.sustainabilityexchange.ac.uk/learning_and_teaching.

Table 1. Overview of Selected Literature: Examples at Cross-institutional and Institutional Levels, as well as Other Supports.

Program/Study/Institution	Details	References
Cross-institutional Level		
Sustainability and Education Policy Network (SEPN) – International	Sustainability Education Research Institute, University of Saskatchewan. Fills knowledge gaps about how education systems influence sustainability. Funded by the Social Sciences and Humanities Research Council.	https://sepn.ca/research_results/sustainability-canadian-post-secondary-institutions-leaders-laggards-live/
50 Canadian PS institutions	41 Strategic plans had inclusion of sustainability; most detailed from institutions participating in STARS (Sustainability Tracking, Assessment and Rating program) (see below).	Bieler and McKenzie (2017)
Association for the Advancement of Sustainability (AASHE)	Supports PS institutions to enhance sustainability practices. 930 institutions have registered to use STARS – majority in the US, plus 30 other countries. In Canada: 10 provinces and territories. BC: 12 institutions.	http://www.aashe.org/
220 accredited institutions in Canada	Assessed leadership on sustainability initiatives. 99 Institutions signed a declaration, e.g., Talloires, a 10-point plan for incorporating sustainability into higher education; 66 of those undertaking a sustainability initiative. Institutions in Quebec (85%) and BC (67%) have most.	Beveridge, McKenzie, Vaughter, and Wright (2015)
70 PS institutions worldwide	Survey found strong linkages between an institution's commitment to sustainability, implementation, and signing a declaration, charter, or initiative.	Lozano et al. (2015)
All Association of Universities and Colleges of Canada (AUCC) member institutions in Atlantic Canada	Most institutions engaged in curricular sustainable development work. Research and scholarship conducted by faculty and students. Research: sustainability questions across campus in collaboration with community partners.	Beringer, Wright, and Malone (2008)
Education for Sustainability Action Group	Educational Developers Caucus, Society for Teaching and Learning in Higher Education. Shared ideas, information, and opportunities related to education for sustainability and its support at the educational developer, instructor, institutional, and national levels.	https://www.stlhe.ca/affiliated-groups/educational-developers-caucus/resources/education-for-sustainability/
UN Sustainable Development Goals (SDGs)	Seventeen global challenges applicable to any course or program, e.g., zero hunger, gender equality, avoiding waste of food or water, climate change, life on land and under water, clean energy and industry, innovation, and infrastructure.	https://www.un.org/sustainabledevelopment/sustainable-development-goals/
National Action Plan for Education for Sustainability, Australian Government	"Living Sustainably" works to integrate education for sustainability into all PS areas.	Commonwealth of Australia (2009)
Develop EAP: A Sustainable Academic English Skills Course	Content based on the UN's SDGs (above). Course development began in 2014; the SDG iteration piloted at the University of Macau in 2016. Free download has been accessed by nearly 300 institutions.	Bolster and Levrai (2017) https://developeap.weebly.com/
Institutional Level		
University of British Columbia (UBC)	First to adopt a sustainable development policy (1997) and to open sustainability office (1998). Twenty-year Sustainability Strategy (2014); Shaping UBC's next century plan (2018–2028). Over 600 undergraduate and graduate courses involving sustainability in a searchable database. Community of Practice: Sustainability Across the Curriculum Working Group. Kit with tools to establish and maintain a practical and sustainable ethical platform for international engagement. Free online, sustainability case studies.	www.sustain.ubc.ca/ https://sustain.ubc.ca/teaching-applied-learning/courses Cassidy (2011a) Ethics of International Engagement and Service-Learning Project (2011)

	CSL project assignment promoted critical reflection in a third year science elective. Students responded to a real need in the community, studying course topics while also practising skills applicable to any discipline: scholarly research, field work, networking, communications, and creativity.	Cassidy (2011b)
University of Guelph	Lists 260 sustainability-related courses and student sustainability research projects.	https://www.uoguelph.ca/sustainability/our-progress
University of Newcastle, Australia	Suggested auditing and reporting system for embedding sustainability learning across the curriculum. Reaching the conceptual stage, the outcome was hoped to increase disciplinary applications of sustainability.	Date-Huxtable, Ellem, and Robert (2013)
George Brown College	Sustainability-Across-the-Curriculum Audit in response to Ontario government, together with industry, emphasizing the need for graduates to have social, environmental, and economic sustainability skills and knowledge. Made sustainability themes visible within courses. That 85% of programs contain at least one learning outcome related to sustainability reflects the college's commitment and position as a system leader. Now increasing the presence of sustainability themes in the curriculum.	Neill (2015) S. Neil (personal communication)
UK	Examples from business, communications, drama, engineering, geography, languages, law, nursing, social work, and theology.	Jones, Selby, and Sterling (2010)
Simon Fraser University	Designed a new undergraduate program in sustainable energy engineering. Interdisciplinarity and experiential learning integrated with the goal of creating global citizens who can provide technical leadership in the energy sector.	Neal et al. (2018)
	Sustainability and broader topics of environmental education offered through summer credit field courses, a teacher education program and a two-year Eco Education graduate program.	D. Zandvliet (personal communication)
Royal Roads University	MA in Environmental Education and Communications includes sustainability topics. Students' projects and theses involve curriculum design and development in sustainability, resource management, and conservation.	M. McClaren (personal communication)
Other supports		
University of Guelph	2010 – President's Task Force on Sustainability made recommendations for embedding sustainability in the curriculum. 2013 – Sustainability Learning Fund launched for instructors to incorporate sustainability into existing courses.	https://news.uoguelph.ca/2013/04/new-fund-supports-teaching-sustainability/
Elon University Sustainability Scholars Program	2010 – Grants to undergraduate students for independent, faculty-mentored sustainability research.	https://www.elon.edu/u/bft/sustainability/get-involved/student-opportunities/sustainability-research-scholars/
University of Pennsylvania Integrating Sustainability Across the Curriculum (ISAC) Program	Undergraduate research assistants and instructors incorporate sustainability into courses.	Dmochowski, Garofalo, Fisher, Greene, and Gambogi (2016)
McGill University Office of Sustainability	Resources, events and opportunities for students, staff, and faculty. Sustainability Projects Fund, the largest in Canada, open to the university community. Projects completed in education, research, operations, connectivity, and governance and administration.	https://www.mcgill.ca/sustainability/about https://mcgill.ca/sustainability/spf/projects/all-spf-projects

EDUCATIONAL DEVELOPMENT FOR SUSTAINABILITY

Examples of direct support for those teaching in post-secondary education are sparse. We have previously described a handful of workshops to introduce or enhance sustainability into courses or programs (Cassidy et al., 2014), including the well-known US Ponderosa, Piedmont, and Penstemon Projects, all designed for faculty across the disciplines, the Pacific Spirit Project at the British Columbia Institute for Technology (BCIT) for faculty within the School of Construction and the Environment, and also in BC, the Sustainability Education across the Province (SEAP) – designed by two student participants as their SEI project – offered to faculty teaching credit courses.

We have identified four other workshops at post-secondary institutions:

- At Tecnologico de Monterrey (Institute for Technology and Higher Education; ITESM) Monterrey campus, Mexico, a team of faculty members, working with the campus sustainability coordinator, designed a three-day Educate the Educator course in 2003 about sustainable development (Lozano-García et al., 2008), with up to 100 participants per course. Connected to Tec's mission and vision statements, requiring that the concept of sustainable development be incorporated into professional education and institutional activities, the multidisciplinarity of instructors who taught the course was found to contribute to its effectiveness.
- At Thompson Rivers University in BC, the Centre for Excellence in Learning and Teaching and the Office of Sustainability offer a two-day workshop to train faculty to integrate sustainability into their curriculum https://www.tru.ca/celt/faculty-learning/sustainability-in-the-curriculum.html; MacEwan University in Alberta has also offered this workshop.
- Royal Roads University in BC has begun a three-year initiative bringing together seven post-secondary institutions across BC and six industry associations. Inspiring Climate Action http://resiliencebydesign.com/, funded by Natural Resources Canada and BC Climate Action Secretariat, designs, develops, and delivers curricula for working professionals, expanding skills, and knowledge to increase BC's capacity to adapt to climate change (V. Forssman, personal communication).
- The Centre for Pedagogical Innovation at Brock University in Ontario offers Promoting Sustainable Development within the Classroom as a Teaching Assistant Workshop, a component of its Certificate Program https://brocku.ca/pedagogical-innovation/what-we-offer/instructional-development/certificate-in-teaching-and-learning-in-higher-education/. Using the SDGs noted in Table 1, this workshop provides examples and suggestions on how to implement them within the classroom and greater community. Initiated and delivered by Nico Gadea, a graduate student, it is the only graduate level workshop in sustainability training and implementation that we have found.

A handful of commercial enterprises relate to "train the trainer" for sustainable education. In terms of other ways to share ideas and resources about sustainability for educators, Algonquin College in Ontario offers a part-time online certificate in Sustainability Education https://www.algonquincollege.com/ccol/program/

sustainability-education-part-time-online/. Neal, Cassidy, and Macpherson (2013) shared examples from Canada, the UK, and Australia.

The United Nations Educational, Scientific and Cultural Organization (UNESCO) offers materials, with activities and resources, including Teacher Education for a Sustainable Future and Sustainable Development across the Curriculum http://www.unesco.org/education/tlsf/mods/theme_b/mod06.html.

Starting in 2015, Neal, Cassidy, and Neill (2017) and Neal, Neill, and Cassidy (2017), interviewed 10 post-secondary educational developers across Canada, asking them to describe how they support Education for Sustainability (EfS). The resulting videos and summary transcripts (http://www.sfu.ca/tlc/resources-and-tools/sustainability.html) demonstrate a wide diversity of ways that educational developers contribute to EfS, at the national, local, institutional, faculty/department, and course levels. They support post-secondary teachers to include sustainability themes, pedagogies, outcomes, and objectives into their courses and programs.

For example, teaching design at the University of Prince Edward Island, Libby Osgood and colleagues divided aspects of sustainability into modules to share content expertise across semesters, bringing more attention to social sustainability. Heather Ranson notes a dedicated, required sustainability course in the B.Comm and MBA programs at the University of Victoria, BC and showcases faculty who have written sustainability business cases.

When an educational developer at two Ontario universities, Western, and Guelph, Gavan Watson supported faculty through funding and a community of practice. Trish O'Connor and Jane Gray (2016) make sustainability a key strategy campus-wide at Fleming College, Ontario. A web resource library gives faculty easy access to materials, including a guide to help them integrate sustainability into a course and workshops supporting discussion across disciplines.

Other teacher-support programs, from Canada, the US and the UK are listed at https://cassidyinview.wordpress.com/by-topic/sustainability/sustainability-education-and-leadership/sustainability-programs/including examples specific to Ontario and BC (Neal, Watson, Neill, & Cassidy, 2014).

SUSTAINABILITY EDUCATION INTENSIVE

Overview

We created a short course, called the SEI, and have previously detailed its major components, including steps in planning, design, promotion, and delivery such that others could use or adapt it at their institution (Cassidy et al., 2014). The goal of the SEI was to provide tools, resources, structured time, and support for participants to develop a Sustainability Education Plan (SEP) to integrate or enhance sustainability in an educational project, be it a credit course, a credit or non-credit program, or another kind of initiative in a formal or non-formal setting.

The SEI was unique and the first of its kind for a variety of reasons. Designed to mirror a unique aspect of the UBC – that people in a variety of roles are active leaders in sustainability education – the SEI was designed and facilitated by a faculty member (Cassidy), a graduate student (Sipos), and an undergraduate student (Nyrose). People in all of these roles, and from a wide variety of academic disciplines, were invited to apply.

Through the application process, or pre-program inquiry, participants described their teaching, facilitation, and/or leadership activities and their ideas for change in a course, program, or initiative. These ideas formed the basis for a SEP, which they worked on throughout the workshop, honing it with resources and discussion. Participants were further supported through studio sessions and professional development funds. They returned a year later to present the results of their work at a Showcase event.

For the time and energy that participants put into applying for, taking part in and following up, we wanted to summarize the workshop aspects they found helpful, and how participation affected them as sustainability educators. We sought insights into how educators can be supported through the process of integration of sustainability into curricular and co-curricular learning experiences.

Methods

We used an ethics protocol for human subjects approved by the Behavioral Research Ethics Board at UBC (BREB #H09-00824) for 30 individuals who gave consent and who were participating in the SEI. We coded the data, which was aggregated and anonymized, identifying emergent themes. We collected the following kinds of materials:

- *Application (pre-program inquiry):* Described above, this also served as an inquiry into participants' expectation of and interest in the program.
- *Formative feedback:* Written feedback on each Days 1 and 2 of each workshop, was summarized and presented the following day, giving voice to participants and allowing us to make adjustments to best meet their needs.
- *SEP:* Considered to be "working documents," SEPs were created by participants for their own personal use and honed throughout the workshop. They received feedback from facilitators and fellow participants. They presented highlights in the form of mini-posters during the course, brief summary reports of how they used their professional development funds, and through sharing during the Showcase event.
- *Summative (post-program inquiry):* Administered on Day 3, we gave participants up to two weeks to reply, asking them about their learning process, and providing an opportunity for general feedback.
- *Post-program inquiry (one year later):* We asked participants many of the same questions as in the summative, again with up to two weeks to reply.

Data for the two post-program inquiries were collected via the Lime Survey online tool.

All forms and questionnaires, as well as some summaries noted below are found at https://cassidyinview.wordpress.com/teaching-and-learning-workshops/sustainability-education-intensive-a-workshop-for-post-secondary-teachers/.

RESULTS

A total of 36 participants took part in an SEI over its two-year run, of whom 30 provided consent (for some questions, if fewer than 30 responded, we present sample size).

Application (pre-program inquiry): Roles included faculty (14), including visiting scholar, assistant professor, senior instructor, assistant dean, instructor, curriculum design specialist and adjust professor; undergraduate student (7); graduate student (5); staff (4); and researcher (3). Because some people held more than one role, numbers do not add up to 30. The range of years in their role ranged from 0.5 to 20, averaging 4.3 years. Not surprisingly, the amount of time that students (both undergraduate and graduate) had been in their roles lower (0.5–2 years), with faculty longer (3–20 years).

Participants noted interest in introducing or enhancing aspects of sustainability in credit courses (from 1st through 4th year and graduate courses), ranging in class size – less than 20 (four such courses), 20–50 (three courses), 50–100 (one course), 101–250 (four courses), greater than 250 (three courses). Non-credit programs and initiatives were the focus for 10 participants, including one campus-wide initiative, one graduate student conference, one certificate program and several non-credit courses and other programs and initiatives. Even at its most conservative, an estimated 1,500 students could benefit. Clearly there was a great potential impact of the two years' cohorts on sustainability across the curriculum and more broadly across several aspects of academic life.

Formative feedback: We present the aggregate summary of formative feedback at https://cassidyinview.wordpress.com/teaching-and-learning-workshops/sustainability-education-intensive-a-workshop-for-post-secondary-teachers/ (scroll down to Outcomes).

Participants expressed many very positive comments. They enjoyed the settings and activities, learned much, found useful resources and hence looked forward to returning on each subsequent day. They especially enjoyed talking with and learning from fellow participants in their diverse roles, stages, and disciplines. Suggestions for change included making more time for some activities and discussions, preferring depth rather than breadth in some cases. One activity we introduced in Year 2 – a web-based resource scavenger hunt– was not well-received, because of set-up time, technical glitches on our part, and, for some participants, lack of technology skill.

SEP: Mirroring their applications (pre-program inquiries), SEP contexts ranged from the micro- to macro-level, with examples of individual credit courses, non-credit programs, or initiatives; across an academic program, department, or faculty; and institution- or organization-wide.

Though the majority of SEPs were at UBC, nine other institutions or organizations were the focus of one or more SEPs, including at two institutions in Ontario, two in British Columbia, and one in China.

The full lists of SEP titles and highlights are at https://cassidyinview.wordpress.com/teaching-and-learning-workshops/sustainability-education-intensive-a-workshop-for-post-secondary-teachers/ (scroll down to Outcomes).

SUMMATIVE (POST-PROGRAM INQUIRY, DAY 3) AND POST-PROGRAM INQUIRY (ONE YEAR LATER)

In responding to the question "What did we do that helped you learn?" participants were enthusiastic about most of the suggested items listed on the survey, with 70% or more of each item being noted as helping their learning (see Table 2). Two items scored lower one year later than immediately after the workshop. Perhaps the "other" items that scored higher one year later were more memorable.

Open-ended responses to the question "Why was this aspect valuable?" included the terms support, encouragement, inspiration, ideas, and energy. Participants appreciated the diversity of academic backgrounds and perspectives, as well as the various tools, techniques, strategies, and resources. Many respondents also referred to time to reflect on and work on their SEPs and receive feedback from peers, and the outdoor activities that "they saw with new eyes." Examples of Day 3 open-ended comments:

> Although sustainability has been a principle I have supported for years, it was good to be in an environment where there was a sharing of methods and strategies used in the classroom. I made some excellent contacts.

> The networking and support helps inspire and move you along. It is easy to get discouraged when you feel you are having to figure everything out by yourself. Sometimes I felt like I'm alone – now I know I'm not.

> It is amazing that a variety of project ideas from individuals of different backgrounds really have a lot in common, not just the idea of sustainability. It spurs even more ideas, methodologies and know-how transfer techniques, which I find very useful and effort/time-saving. Gratifying and inspiring.

Table 2. Responses to a Key Learning Process Question on Day 3 of SEI and One Year after SEI: "What Did We Do That Helped You Learn? Check Off All That Apply" Day 3: $N = 28$ participants; 1 year later $N = 18$ participants. Percentage of occurrence for each response, averaged between 2009 and 2010. (People selected more than one item listed, so percentages in a given column add up to more than 100%. When choosing Other, participants wrote in a text box.)

Day 3 (%)	Item Listed	1 Year Later (%)
82	Becoming part of a face to face network of educators interested in sustainability education.	68
79	Introducing a variety of definitions for, and perspectives about sustainability and sustainability education.	49
70	Having structured space and time to work on and reflect on enhancing your course, program, or initiative.	80
79	Experiencing and reviewing various tools and techniques that contribute to sustainability education.	49
88	Accessing resources related to sustainability education.	70
14	Other (examples) • First day at UBC Farm. • Field trip around campus. • Funding to support SEP. • Participating in showcase event.	47

In responding to the question "Do you think you have changed as a sustainability educator through your participation in the SEI?" on Day 3 ($N = 28$ participants responding), 86% said Yes; one year later ($N = 18$ participants responding), 58% said Yes, with the remainders saying No, Unsure or not responding. Participants further explained:

Examples of "Yes":

> Increased self confidence that I am on the right foot to continue involving people into changing behaviors through education in sustainable practices.

> I have more of a clear idea of where to start. My SEP is a small but necessary step to get me more oriented.

Examples of "No" or "Unsure":

> Having seen the effort at UBC, I now have a lot of ideas for my own plans.

> I appreciate the difficulties more.

> I don't know yet. Maybe. I think I'll think more about how to enhance environmental sustainability.

Perhaps those who responded "No" or "Unsure" were still processing what they had experienced. It would be interesting to ask again two or more years later.

In the post-program survey (one year after participants took part in an SEI), in responding to the question "Have you experienced shifts or changes in your knowledge, skills and attitudes related to sustainability?" ($N = 14$), 65% said yes. Participants further explained:

> I have become a more effective sustainability educator, through incorporating new approaches and activities in my teaching. My commitment to sustainability education has also strengthened through participating in the SEI.

> I have a better understanding of the underlying theory and of where to look for reliable resources. I feel more connected to the sustainability efforts on campus, therefore more confident in my personal activities.

> I think more about environmental sustainability in educational programs I am planning and have changed my own habits.

DISCUSSION

Participants put much effort into all stages of the SEI workshop. Their SEPs were a key focus throughout, from application to follow-up. Their feedback, from formative, Day 3 and one year later, indicates to us that the workshop was valuable to them. It is not surprising that survey response sample sizes went down between Day 3 and one year later, with these faculty, staff, and students all juggling many commitments. Though low sample sizes preclude statistical analyses, open-ended comments indicate that participants continue to reflect on, and make use of the skills, knowledge, and attitudes they developed or enhanced through their participation in the SEI. Additionally, Cassidy et al. (2014) note successes of participants such as through conference sessions and publications.

Though it was beyond the scope of our study to look at effects on students and other recipients of learning by SEI graduates, it would be most interesting to do so.

Kagawa (2007) conducted an online survey to study students' perceptions and understandings of, and attitudes toward sustainable development and related topics. Though a majority of students thought sustainability was "good," this was not strongly related to their understanding of the concepts. She suggested that we help teachers make interconnections more clear, and help students take action. In another study surveying students about sustainability, Emanual and Adams (2011) found that students' views toward the value of sustainability was tied to their understanding of the topic. Through teaching a Human Ecology course for many years, one of us (Cassidy) found that individual students' scores on a self-administered ecological footprint survey (see Wackernagel & Rees, 1996), increased between the start and the end of term. Students mused that this could be a result of learning more about sustainability during term, and hence, being more realistic as the course ended.

An inventory of the in-class activities and assignments that promote students' understanding of and involvement of sustainability could be another way to document effects of a workshop for educators. Moore (2005) argues that educators need to find ways to practice the ideals of sustainability in classrooms, by modeling approaches and considering process as well as content. She suggests, "that the pedagogy of sustainability education is about creating spaces where disciplines are integrated in new ways" (p. 80). Zeegers and Clark (2014) had students keep reflective journals as part of a course focusing on the three pillars of sustainability. Cassidy, Sipos, and Nyrose (2015) describe three activities – sustainability literacy, sustainability "artifact" and "give one, get one" that can be used or adapted in a course or workshop.

How does one go about measuring sustainability in the curriculum? The Environmental Association for Universities and Colleges Scotland studied this question in 2018. Goal 4 of the UN's SDGs, to ensure quality education for all, is echoed by the Scottish Funding Council, with many examples noted https://www.sustainabilityexchange.ac.uk/sustainability_curriculum. The National Union of Students in the UK, which has conducted online surveys since 2011, with 12,000 responses to date; found that 87% of student respondents want their institutions to be doing more on sustainability, with 60% wanting to learn more about it in a course.

CONCLUSION

We provided an overview of selected literature on integration of curricular sustainability development and education and related institutional leadership at the post-secondary level. Turning to educational development for sustainability education, we described various supports, finding very few examples of workshops to help faculty and others who teach.

We described such a workshop, the SEI that we designed and facilitated at UBC for two years. We summarized the workshop aspects that two years of participants found helpful, and how workshop involvement affected them as sustainability educators. We encourage post-secondary institutions to provide support in the form of workshops, tools, resources and funding to help educators introduce or enhance aspects of sustainability into their courses and programs. Students are asking for this, and, as they are future leaders, it is important that educators address the environmental, social and economic issues that demand attention.

ACKNOWLEDGMENTS

We are grateful for financial support of the 2009 SEI from the Sustainability Office and the then-named Centre for Teaching and Academic Growth at the University of British Columbia (UBC). Funding for the 2010 SEI came from the Teaching and Learning Enhancement Fund made possible by UBC students.

We are grateful to the Educational Developers Caucus (EDC) of the Society for Teaching and Learning in Higher Education for funding two projects. In 2012, Resource Repository for Sustainability Education and Leadership was granted to Cassidy, hiring Braydon Walker when a third year undergraduate student, Engineering, Carleton University and Sipos when a graduate student at UBC. In 2015, Best Practice Cases of Education for Sustainability in Educational Development was granted to Vivian Neal, Simon Fraser University (SFU); Sandra Neill, George Brown College; and Cassidy (UBC), hiring Research Assistant Kar-Bo Ng when an undergraduate student at UBC, work-study students Laura Guevara and Junone Kang, and videographer Duane Woods at SFU.

Thanks to EDC colleagues who responded to our listserv request for additional examples: Carolyn Ives, Thompson Rivers University; Nicholas Gadea and Lianne Fisher, Brock University; Vivian Forssman, Royal Roads University; and Milt McClaren and David Zandvliet, SFU.

REFERENCES

Beasy, K. M., Peterson, C. J., Tomlinson, A., & Tiernan, B. (2016). Curriculum for the future: Sustainability is a must! *Teaching Matters Conference Programme*, 7 December, Hobart, Tasmania, Australia (p. 8).

Beringer, A., Wright, T., & Malone, L. (2008). Sustainability in higher education in Atlantic Canada. *International Journal of Sustainability in Higher Education*, *9*(1), 48–67. https://doi.org/10.1108/14676370810842184

Beveridge, D., McKenzie, M., Vaughter, P., & Wright, T. (2015). Sustainability in Canadian post-secondary institutions. *International Journal of Sustainability in Higher Education*, *16*(5), 611–638. http://dx.doi.org/10.1108/IJSHE-03-2014-0048

Bieler, A., & McKenzie, M. (2017). Strategic planning for sustainability in Canadian higher education. *Sustainability*, *9*(2), 161. doi:10.3390/su9020161

Bolster, A., & Levrai, P. (2017). A slow (r)evolution: Developing a sustainable EGAP course. *The European Journal of Applied Linguistics and TEFL*, *6*(1), 147–165.

Brinkhurst, M., Rose, P., Maurice, G., & Ackerman, J. D. (2011). Achieving campus sustainability: Top-down, bottom-up, or neither? *International Journal of Sustainability in Higher Education*, *12*(4), 338–354. doi:10.1108/14676371111168269

Cassidy, A. (2011a). Communities of practice: A checklist for success. *Transformative Dialogues: Teaching and Learning eJournal, 4*(3). Retrieved from http://www.kpu.ca/sites/default/files/Teaching%20and%20Learning/TD.4.3.6_Cassidy_Communities_of_Practice.pdf

Cassidy, A. (2011b). Building critical reflection into CSL Group assignments: Show what you learned in school today. *Global Praxis: Exploring the ethics of engagement abroad* (pp. 198–202). Vancouver, BC. Retrieved from https://cassidyinview.wordpress.com/my-publications/specific-to-biology-345-human-ecology/

Cassidy, A., & Sipos, Y. (2007). Students as co-researchers and co-presenters. Presented at the 27th annual conference of the society for teaching and learning in higher education (STLHE) – Evolving Scholarship, June 13–16, University of Alberta, Edmonton, AB. Retrieved from https://cassidyinview.wordpress.com/by-topic/sustainability/sustainability-education-and-leadership/peer-reviewed-conference-sessions/

Cassidy, A., Sipos, Y., & Nyrose, S. (2014). Supporting sustainability education and leadership: Strategies for students, faculty, and the planet. In S. Mukerji & P. Tripathi (Eds.), *Sustainable practices: Concepts, methodologies, tools, and applications* (pp. 1454–1477). Hershey, PA: Information Science Reference. doi:10.4018/978-1-4666-4852-4.ch081. Retrieved from https://www.igi-global.com/chapter/supporting-sustainability-education-leadership/78122; https://open.library.ubc.ca/cIRcle/collections/ubccommunityandpartnerspublicati/52387/items/1.0376045

Cassidy, A., Sipos, Y., & Nyrose, S. (2015, September). Sustainability education: Three activities you might use or adapt to help educators. *Transformative Dialogues: Teaching and Learning eJournal, 8*(2). Retrieved from http://www.kpu.ca/sites/default/files/Transformative%20Dialogues/TD.8.2.4_Cassidy_etal_Three_Activities_to_Help_Educators.pdf

Commonwealth of Australia. (2009). *Living sustainably: The Australian government's national action plan for education for sustainability*. Canberra: Department of the Environment, Water, Heritage and the Arts.

Date-Huxtable, E., Ellem, G., & Robert, T. (2013). The low carbon curriculum at the University of Newcastle, Australia. In S. Caeiro, W. Leal Filho, C. Jabbour, & U. M. Azeiteiro (Eds.), *Sustainability assessment tools in higher education institutions* (pp. 345–357). Switzerland: Springer International Publishing. https://doi.org/10.1007/978-3-319-02375-5_19

Dmochowski, J. E., Garofalo, D., Fisher, S., Greene, A., & Gambogi, D. (2016). Integrating sustainability across the university curriculum. *International Journal of Sustainability in Higher Education, 17*(5), 652–670. https://doi.org/10.1108/IJSHE-10-2014-0154

Emanual, R., & Adams, J. N. (2011). College students' perceptions of campus sustainability. *International Journal of Sustainability in Higher Education, 12*(1), 79–92. https://doi.org/10.1108/14676371111098320

Ethics of International Engagement and Service-Learning Project. (2011). *Global praxis: Exploring the ethics of engagement abroad*. Vancouver, BC. Retrieved from http://ethicsofisl.ubc.ca/

Filho, W. L., Wu, Y.-C. J., Brandli, L. L., Avila, L.V., Azeiteiro, U. M., Caeiro, S., & Madruga, L. R. R. G. (2017). Identifying and overcoming obstacles to the implementation of sustainable development at universities. *Journal of Integrative Environmental Sciences, 14*(1), 93–108.

Jones, P., Selby, D., & Sterling, S. (2010). *Sustainability education: Perspectives and practice across higher education*. Abingdon: Earthscan.

Kagawa, F. (2007). Dissonance in students' perceptions of sustainable development and sustainability: Implications for curriculum change. *International Journal of Sustainability in Higher Education, 8*(3), 317–338. https://doi.org/10.1108/14676370710817174

Lefebvre, D. (2000). Sustainability education evaluation tool. In A. Jarnet, B. Jickling, L. Sauve, A. Wals, & P. Clark (Eds.), *A colloquium on the future of environmental education in a postmodern world*. Whitehorse, Canada: Yukon College.

Lozano, R., Ceulemans, K., Alonso-Almeida, M., Huisingh, D., Lozano, F. J., Waas, T., …, Hugé, J. (2015). A review of commitment and implementation of sustainable development in higher education: Results from a worldwide survey. *Journal of Cleaner Production, 108*(A), 1–18.

Lozano-García, F. J., Gándara, G., Perrni, O., Manzano, M., Hernández, D. E., & Huisingh, D. (2008). Capacity building: A course on sustainable development to educate the educators. *International Journal of Sustainability in Higher Education, 9*(3), 257–281.

Lugg, A. (2007). Developing sustainability-literate citizens through outdoor learning: Possibilities for outdoor education in higher education. *Journal of Adventure Education and Outdoor Learning, 7*(2), 97–112.

Maloni, M. J., Smith, S. D., & Napshin, S. (2011). A methodology for building faculty support for the United Nations principles for responsible management education. *Journal of Management Education*, *36*(3), 312–336.

Moore, J. (2005). Is higher education ready for transformative learning? A question explored in the study of sustainability. *Journal of Transformative Education*, *3*, 76–91. https://doi.org/10.1177/1541344604270862

Neal, V., Cassidy, A., & Macpherson, A. (2013). Who? Us? What can EDs do about the really big issues: Global warming and sustainability? *Conference on Educational Developers Caucus (EDC)*, Wilfrid Laurier University, Brantford, ON. Retrieved from https://cassidyinview.wordpress.com/by-topic/sustainability/sustainability-education-and-leadership/peer-reviewed-conference-sessions/

Neal, V., Cassidy, A., & Neill, S. (2017). Cases of education for sustainability in educational development. *Society for teaching and learning in higher education (STLHE) conference*, June, Dalhousie University, Halifax, NS. Retrieved from https://wordpress.com/page/cassidyinview.wordpress.com/1626

Neal, V., Neill, S. C., & Cassidy, A. (2017). Education for sustainability (EfS). Retrieved from http://www.sfu.ca/tlc/resources-and-tools/sustainability.html

Neal, V., Oldknow, K., Edgar, J., Bajić, I., Trautman, M., & Moallem, M. (2018). A new program in sustainable energy engineering: Balancing subject matter with transformative pedagogies to produce global citizens. *Proc. of the 9th Engineering Education for Sustainable Development*, June 3–6, Glasboro, NJ.

Neal, V., Watson, G., Neill, S., & Cassidy, A. (2014). Education for sustainability: Highlights from four institutions. *Conference on Educational Developers Caucus*, University of Calgary, Calgary, AB. Retrieved from https://cassidyinview.wordpress.com/by-topic/sustainability/sustainability-education-and-leadership/peer-reviewed-conference-sessions/

Neill, S. C. (2015, September). Sustainability-across-the-curriculum audit at George Brown College. *Transformative Dialogues: Teaching and Learning Journal*, *8*(2). Retrieved from https://www.kpu.ca/sites/default/files/Transformative%20Dialogues/TD.8.2.7_Neill_Sustainability_Across_the_Curriculum.pdf

O'Connell, T. S., Potter, T. G., Curthoys, L. P., Dyment, J. E., & Cuthbertson, B. (2005). A call for sustainability education in post-secondary outdoor recreation programs. *International Journal of Sustainability in Higher Education*, *6*(1), 81–94. https://doi.org/10.1108/14676370510573159

O'Connor, T., & Gray, J. (2016, February). Incorporating sustainability across the curriculum. *Sustainability: The Journal of Record*, *9*(1). https://doi.org/10.1089/SUS.2016.29037

O'Flaherty, J., & Liddy, M. (2018). The impact of development education and education for sustainable development interventions: A synthesis of the research. *Environmental Education Research*, *24*(7), 1031–1049. https://doi.org/10.1080/13504622.2017.1392484

Wackernagel, M., & Rees, W. (1996). *Our ecological footprint: Reducing human impact on the earth*. Philadelphia, PA: New Society Publishers.

World Commission on Environment and Development. (1987). *Our common future*. Oxford: Oxford University Press.

Zeegers, Y., & Clark, I. F. (2014). Students' perceptions of education for sustainable development. *International Journal of Sustainability in Higher Education*, *15*(2), 242–253. https://doi.org/10.1108/IJSHE-09-2012-0079

CHAPTER 6

HORTUS IN URBE: BUILDING A SUSTAINABLE DEVELOPMENT CURRICULUM IN CHICAGO

Euan Hague, Howard Rosing and Joseph P. Schwieterman

ABSTRACT

In this chapter the authors describe the development of an interdisciplinary graduate program focusing on sustainable urban development at Chicago's DePaul University. Locating the curriculum both in the administrative institutional context and the historical geography of a racially and economically segregated urban area, the authors discuss the process of program formation and adjustment over its first five years of operation, 2013–2018. The chapter highlights some of the challenges encountered by program faculty, from internal curricular competition to external classification of the program under federal educational designation, and notes some of the interdisciplinary innovations, such as requiring courses in Geographic Information Systems to aid spatial data analysis and visualizations. In the second part of the chapter, to assess the impact of the graduate program on students, the authors review and draw from reflection essays written by students who, while completing their studies, pursued internships with nonprofit organizations in Chicago. The chapter concludes by discussing the employment status of recent alumni to suggest how these former students are pursuing jobs that may come to impact sustainability policy and practice. The authors maintain it is necessary to push the understanding of "sustainability" beyond solely environmental concerns

Integrating Sustainable Development into the Curriculum
Innovations in Higher Education Teaching and Learning, Volume 18, 91–105

ISSN: 2055-3641/doi:10.1108/S2055-364120200000018024

to incorporate understandings of how economic development and community engagement must be included to deliver a sustainable city. The interdisciplinary curriculum described challenges students to become leaders in local efforts to make urban areas not only more environmentally sustainable, but also more economically and socially sustainable for all residents.

Keywords: Sustainability; urban development; urban planning; Chicago; curriculum; Geographic Information System; interdisciplinary; internships; graduate students; community; United States

TEACHING URBAN SUSTAINABILITY IN CHICAGO

The mission and evolution of the sustainable urban development (SUD) graduate program at DePaul University in Chicago, Illinois, differs from many graduate degrees in the field of sustainability studies. The program uses its home city as an urban laboratory in which students evaluate city- and neighborhood-oriented development issues with an eye toward finding sustainable solutions to its most pressing problems. These problems encompass Chicago's struggles to help overcome the legacy of racial segregation and as a place experiencing some of the harshest forms of structural inequality in the United States (Shabazz, 2015). Students explore these issues while drawing upon lessons from public policies adopted both domestically and internationally.

Chicago's reputation as a city of sharp contrasts motivated us – the designers of the program – to develop a curriculum that required students to ask difficult questions about the goals, processes, and promises of SUD. The result is a program that leverages the city's complex urban fabric, gives students interdisciplinary, theoretical, and practical learning experiences, and offers a critical vision of sustainable development education in an urban context. This chapter expounds upon the development of our academic offering and concludes with a summary of lessons learned during the program's first six years.

THE CURRICULAR AND URBAN CONTEXT

During the first decade of the 2000s, advocates for advancing sustainability education lamented that traditional disciplinary boundaries were barriers to developing curriculum. In the context of British higher education, for example, Blewitt (2004, p. 2) argued that:

> Sustainability is complex and complicated, with no single discipline definitively addressing either the problems or the solutions: it incorporates technological, philosophical, economic, social, ecological, political and scientific dimensions. This may be illustrated through an examination of real-world issues or projects that are motivated by concerns over sustainability – for example, in Green architecture, eco-design, gender and development; integrated and sustainable transport; global citizenship; and lifelong learning.

Later in the decade, Sterling (2009, p. 106) argued that sustainability was a radical epistemology that necessitated fundamental transformations in higher education, and yet it remained "at the margins of debate and practice." In the UK, Edwards (2004, p. 135) points out, trends since the mid-1990s led to superficial "greening of courses that traditionally had ignored environmental issues." Others, such as Kevany, Huisingh, and Lozano García (2007, p. 103), demanded:

> that educational leaders in all nations must make fundamental changes in the ways they engage their students, fellow-faculty and community stakeholders in helping to chart the course for and in developing people with the capacity to make the transition to sustainable societies.

Blewitt (2004, p. 3) advocated meeting these curricular challenges by overcoming disciplinary inertia and devising new programs that were "multidisciplinary and interdisciplinary learning opportunities as well as practice-based educational experience." Yet, as Jones, Selby, and Sterling (2010, p. 34) highlight:

> Embedding interdisciplinarity into the policies, practices and cultures of [higher education institutions] is difficult because it challenges cultural and structural norms; yet, where it is done well, it affords a renewal of pedagogy and stimulus to learning and a boost to collaboration and cross-fertilization that benefits the host institution – both internally and in terms of its reputation.

Within this context, in 2011 DePaul conducted a curricular review for the Association for the Advancement of Sustainability in Higher Education's (AASHE) Sustainability Tracking, Assessment & Rating System (STARS). This self-study identified a multidisciplinary array of existing sustainability-focused and sustainability-related courses at the undergraduate level, but a dearth at the postgraduate level. Only one of DePaul's 78 graduate programs, a Masters in International Public Service, met the AASHE/STARS guidelines for delivering a sustainability-focused degree. Our charge was, therefore, to devise a graduate program that: (a) drew on existing courses; (b) added critical new curricular offerings; (c) gave students opportunities to put their learning into practice; and (d) would advance sustainable development in metropolitan Chicago. DePaul's graduate students, we believed, needed a curriculum that would enable them to explore sustainable development in the context of Chicago's urban story.

Incorporated in 1837, Chicago is relatively young by international standards. It developed following the 1833 Treaty of Chicago which removed thousands of indigenous peoples to land west of the Mississippi River prior to subsequent resettlement of the area by migrants from the USA's east coast and beyond. Early leaders gave Chicago its motto "Urbs in Horto" – city in a garden – a phrase still displayed on the city's official seal and regularly cited by officials in policy debates. This Latin expression refers to the city's development amid what were, at the time, grassland prairies. After the Great Fire in 1871, Chicago quickly rebuilt and emerged as a manufacturing powerhouse of immigrant communities, skyscrapers, and vast stockyards. By the nineteenth century's end, Chicago was one of the world's largest and most industrialized cities and rising class conflict contributed to the birth of the international labor movement (Abu-Lughod, 1999; Pacyga, 2009). Subsequently, the Great Migration of African Americans to northern US cities in the first half of the twentieth century saw Chicago evolve into what is, in effect, two cities in one: one largely constituting the descendants of white Europeans and another made up of African American residents segregated into what became known as the "Black Belt" (Burgess, 1925; see also Abu-Lughod, 1999; Washington, 2005). Consequently, present-day Chicago suffers from acute racial and class inequalities. Residents in poorer neighborhoods, largely African American and Latinx, struggle with limited access to quality education, transportation, food, healthcare, and employment. Yet within the city's neighborhoods, community organizers, associations, and neighborhood-based organizations remain resilient and, as described below, resist in ways that present challenges to narrowly defined notions of "sustainability," such as those drawn solely from environmentalism.

As the metropolis grew, Chicago's philanthropists and urban planners maintained a commitment to the vision exemplified by its motto. Municipal legislation mandated that the shoreline of Lake Michigan remain "forever open, clear, and free" (Wille, 1991) and city and county officials established a network of parks and forest preserves in the early-twentieth century (Fisher, 2015). More recently, under the lengthy administration of Mayor Richard M. Daley, who served from 1989 to 2011, the city began to advance a commitment to urban development that integrated the concept of sustainability. Styling himself as a "green mayor" and utilizing funds received from utility companies following a deadly 1995 heatwave (Klinenberg, 2015; Lorr, 2017), Daley initiated a series of annual reports that supposedly set a path toward sustainable development for Chicago. The City issued its first *Environmental Action Agenda* report in 2005, calling for environmental conservation and environmentally aware decision-making by both the city's government and its residents (Lorr, 2017).

By the mid-2010s, Chicago had been releasing *Action Agenda* updates annually, touting programs to create greater acreage of green roofs and green alleys, give away rain barrels, promote LEED-standard building construction, and add miles of bike lanes. Daley's successor, Mayor Rahm Emanuel, in office 2011–2019, appointed a Green Ribbon Commission comprising of and led by major corporations to further advance sustainability goals throughout the metropolitan region (Lorr, 2017). It was in this context of municipally driven and corporate-led sustainable development policies that faculty from a range of disciplines at DePaul University came together to develop a graduate program focusing on urban sustainability.

DePaul University is a private higher education institution founded in 1898 by the Vincentian Catholic order. In the 2000s it became the USA's largest Catholic university in terms of student enrollment. Its two campuses are each located in affluent areas of the city: the Lincoln Park Campus is in a neighborhood that is more than 95% white and has a median household income of well over $100,000; and the Loop Campus is in the city's central business district, just blocks from the Chicago Board of Trade. After almost two years of curricular discussions, we determined that the program must take advantage of local resources for thinking about, and promoting development in, a city suffering from deep economic and racial divides. By focusing on the university's mission-driven commitment to support Chicago's under-served communities, the resulting SUD curriculum offers a grassroots-oriented alternative to the city government's neoliberal vision of urban development (Bennett, Garner, & Hague, 2017).

We drew insight from other master's programs with similar missions. University of Pennsylvania's Master of Environmental Sustainability, one of the USA's oldest degrees in the field, has a more global focus than the program we espoused to create but showed both how a curriculum could be deeply rooted in the functioning of cities and how a cohesive degree could emerge from courses taught by a variety of academic disciplines. University of Texas at Austin's Planning Practicum Studio course in the Community and Regional Planning master's program was a template for our Urban Design Lab. Our interest in internships was piqued by favorable testimonials about the University of Illinois at Chicago's success in using them to enhance learning experiences in its Master in Urban Planning and Policy program, although our internship requirement would ultimately place stronger emphasis on social-justice considerations, spatial analysis, and neighborhood-scale projects. The UK's University of Nottingham's Sustainable Urban Design program stood out for its specialized focus on the urban built environment and skill-building modules, the scheduling of which echoed DePaul's calendar of 10-week academic quarters.

Taking advantage of the flexibility offered by the 10-week quarter system, rather than the 15-week courses typical of academic semesters, we decided to expose students to an array of disciplinary perspectives, distributed within and across a 13-course degree. This multidiscipliniarity is indicative of our own backgrounds in urban and cultural geography (Hague), anthropology, community service studies and critical food studies (Rosing), and urban planning, public service and transportation (Schwieterman). We believed that students need to learn to integrate economic, environmental, and social notions of sustainability in complicated urban contexts, and proposed a curriculum that sought to overcome the "silos" of academic disciplines that all-to-often result in the separation of these frameworks.

The resulting curriculum puts great emphasis on imparting perspectives on Chicago's social, cultural, political and economic fabric, social justice, and service to communities. It foregrounds technical and critical thinking skills and challenges students to reimagine urban areas and propose new visions for future urban development that integrate practical proposals and achievable goals. Rather than focusing on business-led and consumption-driven solutions, we advance understanding of topics ranging from urban food systems and food justice, alternative transportation, affordable housing, and redesigning urban green spaces for equity and community uses. In this way, students grapple with the messiness and complexity of both sustainability and policy-making. For example, leveraging DePaul's location, which is a short bus or train ride to a remarkably diverse set of neighborhoods, each with complicated histories and hardships, students can explore how housing strategies designed to promote economic development are affected by commercial pressures and transportation needs, and are challenged to understand the problems that occur when sustainability is evaluated solely on the basis of "greening" and enhancing the natural environment.

THE CURRICULUM AND ITS DEVELOPMENT

At first glance, DePaul may appear ideally situated for teaching SUD through a multidisciplinary lens. The university has 10 colleges and schools, several with much coursework related to cities. Yet this same multicollege structure presented challenges as we began to negotiate university-wide curricular options. For example, while devising the SUD graduate program, the College of Business concomitantly developed a Master of Science in Sustainability Management. In conjunction, the Department of Environmental Science and Studies, housed in a College of Science and Health, inaugurated its first graduate-level course, *Earth Resources and Human Society*. Even within our home unit, the College of Liberal Arts and Social Sciences, existing internationally focused graduate courses titled *Sustainable International Development* and *Nature, Society and Power* created a crowded field, and prompted the College's administration to advise a primarily domestic focus for new sustainability-related courses. Such stories of institutional claims to curricular turf are myriad, yet they motivated us to devise a curriculum that would deliver a distinctive, flexible, and collaborative interdisciplinary program while giving students the skills necessary to engage with multiple actors to build sustainable cities.

The program that emerged balances theory with practical application, technology, and professional preparation. That balance gives students a broad base from which to participate in debates about urban sustainability while, as noted in the subsequent sections of this chapter, being able to compete for jobs in a wide variety of related fields. The resulting SUD program came to be seen as a centerpiece of DePaul's urban mission and centered on six required core courses (Table 1). In addition to these core courses, students complete a pair of Geographic Information Systems (GIS) courses and a statistics course. The four remaining courses are drawn from a list of interdisciplinary electives offered by, among others, the university's faculty in Environmental Science and Studies, Food Studies, International Studies, Public Service, Real Estate, and Writing, Rhetoric and Discourse, the latter emphasizing grant writing.

In an effort to move away from traditional methods of assessing student learning, such as papers and exams, faculty are encouraged to develop assignments that reflect tangible skills learned in the six core courses. SUD 401, the introductory course is arguably the most traditional, taught as a seminar with readings and a final paper. In SUD 402, however, students complete a final project using R-Studio software. Typically taken next, SUD 403 requires students to complete a mock application to the Environmental Protection Agency (EPA); SUD 404 asks students to develop a portfolio of spatial visualizations; SUD 501 culminates in a public session in which students present on an innovative or emerging sustainability topic and then publish a short paper online on that topic; lastly, SUD 502 students collaborate on neighborhood projects and then present their findings publicly. Typically, these courses asked students to engage with Chicago's social, political, and cultural context.

From the outset, the program incorporated a 100-hour internship requirement as part of SUD 501. Originally meeting five times per quarter (rather than the 10 sessions common to other courses), and covering a limited range of issues related to professional development, it soon became apparent that both the

Table 1. Summary of Core Courses in Sustainable Urban Development.

Number	Title	Description
SUD 401	Sustainable Urban Development 1: Planning, policy and community development	This course is an intensive, topics-based overview of sustainable land use, housing, transportation, infrastructure, and community and economic development practices. The course is largely directed at contemporary policy issues and technical processes within the United States.
SUD 402	Sustainable Urban Development 2: Green infrastructure and the environment	This course focuses on the environmental aspects of city planning, including: resource management, environmental policy, transportation, LEED ratings and measures of environmental impact, integration of ecosystems within a land use planning framework, and community-based solutions to historical environmental injustice.
SUD 403	Brownfields Remediation: Practice and implementation	This course explores the "brownfield lifecycle," how and why brownfields were created. Through a practice component, class members work in teams to apply their general brownfields knowledge to particular challenges in the field and gain hands-on experience in applied environmental and economic development.
SUD 404	Urban Design Laboratory	This course teaches digital methods of creating visual information. Software used may include: Adobe Creative Suite, Google SketchUp, and advanced techniques for Microsoft PowerPoint. The final project culminates in a "pin up" session, in which students present a digital 3D model of a neighborhood development.
SUD 501	Professional Seminar in Sustainable Urban Development	This course seeks to develop professional, communication, and consultancy skills and explores transdisciplinary approaches to urban development. Speakers interact with practitioners through guest lectures and site visits, culminating in a symposium in which students present to a public audience. Students must complete 100 hours of internship or approved off-campus work.
SUD 502	Capstone – Professional Portfolio	This course is a seminar in which students apply the skills they have learned to a neighborhood-oriented project developed and directed by the instructor. Students produce a comprehensive project report.

structure and content of the course was less-than-optimal for many students. Many found it necessary to accept unpaid internships while simultaneously paying several thousand dollars to take the course. We had expected that some students would complete the internship requirement through their existing employer, for example, by devising a special project related to urban sustainability. Instead, students often requested waivers to replace the internship course or proposed projects that were only tenuously about sustainability in cities. Understanding the importance of professional development for graduates, a major change occurred three years after the program's inception: SUD 501 was separated from the internship requirement, revamped, and renamed *Professional Seminar in Sustainable Urban Development*, a course that spans the full academic quarter.

Redesigning SUD 501 prompted faculty to consider an important question: how can we best use this course to prepare leaders to push cities to develop sustainably? The course, as a result, now puts greater emphasis on *leadership* and is almost always taken by students nearing graduation. During each class session, students engage with urban planners, policy-makers, community advocates, and other sustainability practitioners who deliver guest lectures; some class sessions are scheduled off-campus and hosted by organizations working on urban sustainability issues. These characteristics expand students' professional networking skills and opportunities for employment. Students learn to ask questions, for example, about how racial, gender, and economic equity impact structures of, and daily work on, urban development. Drawing on these experiences, students begin to think more closely about their interests before turning their attention to a final report that is published on either their LinkedIn page or a professional blog. To augment this, students receive training in oral history methods and conduct two interviews with sustainability leaders. They also take part in a colloquium that involves a public presentation. All class material is archived on the course website, including the recorded final presentations, oral histories, reflection blogs, and resumes, making them available to subsequent students and alums.

PROFILE OF STUDENTS

The program adheres to a cohort model, in which most enrolled students start in the autumn term. Most cohorts encompass 12–20 students. Students can nonetheless begin midway through the year, which helps make the program accessible to a diverse group of applicants. The first cohort began during fall of 2013 and the first graduating class of students was in 2015. By January 2019, 55 students had successfully completed the program, 25 of whom identified as men and 30 identified as women. Three students identified as Asian American, three as African American, five as international, 26 as white, and 18 chose not to specify race or ethnicity on the graduation survey. Reflecting the interdisciplinary nature of the program, students came from a wide range of undergraduate backgrounds including the humanities, business, public policy, and urban planning, resulting in a student body with diverse geographic, cultural, and academic backgrounds, which helps foster interdisciplinary learning.

Because of our program's multidisciplinary focus, the program initially faced a classification challenge. Following the program's launch, we realized the importance of the Classification of Instructional Programs (CIP) code, a US Federal Department of Education categorization of degree programs used in assessing numerous administrative aspects and trends in national data. At its inception, DePaul applied SUD the CIP code for Urban Studies/Affairs (45.1201). This coding designates:

> A program that focuses on the application of social science principles to the study of urban institutions and the forces influencing urban social and political life. Includes instruction in urban theory, the development and evolution of urban areas, urban sociology, principles of

> urban and social planning, and the politics and economics of urban government and services. (United States National Center for Education Statistics, 2010a)

We were unaware of this designation by institutional administrators until learning from a colleague that one of the most important elements of a CIP code is that for international students it mandates the duration of the Optional Practical Training (OPT) period postgraduation. Following review of the code descriptions, faculty determined that our curriculum and program requirements more closely aligned with the 30.3301 CIP code for "Sustainability Studies." This coding designates:

> A program that focuses on the concept of sustainability from an interdisciplinary perspective. Includes instruction in sustainable development, environmental policies, ethics, ecology, landscape architecture, city and regional planning, economics, natural resources, sociology, and anthropology. (United States National Center for Education Statistics, 2010b)

Faculty successfully petitioned DePaul's administrators of academic affairs to reclassify the program as "Sustainability Studies." This, in turn, meant that the degree qualified as a federally listed Science, Technology, Engineering, and Math program. One upside is that international students pursuing the degree can, under student visa regulations at the time of writing, remain in the United States in a related job or internship for 24 months after graduation through an OPT extension (United States Department of Homeland Security, n.d.-a, n.d.-b). Under the "Urban Studies/Affairs" designation, this postgraduation period is limited to 12 months.

A related issue was the higher-than-expected number of international applicants, many of whom expressed excitement about the prospect of studying in Chicago but had never been to the city. We felt that the program's orientation toward North American cities – and Chicago in particular – would be problematic for them. Nevertheless, this concern proved unwarranted. International students generally found their acquired skills to be transferrable to overseas locations, while others opted to stay and gain employment locally. In the end, students from different cultural backgrounds abetted meaningful class discussions, as students shared ideas and examples from their home towns, cities, or countries.

STUDYING URBAN SUSTAINABILITY IN CHICAGO

Teaching SUD in a racially segregated city like Chicago offers numerous opportunities to explore why the concept of sustainability must be treated differently across various urban contexts. Most students are struck by the starkly different racial and economic inequalities across Chicago, making it obvious to them that not every neighborhood lives up to the city's "Urbs in Horto" moniker. Lincoln Park is a case in point, having some of the most beautiful public and private gardens in the city, reflected in an annual garden walk event that culminates in a festival on DePaul's campus. Nevertheless, a sizeable percentage of the city's other neighborhoods lack adequate green space (Fisher, 2015; Washington, 2005). Furthermore, students see first-hand that Chicago is a place where municipally driven and corporate-led sustainability and development initiatives tend to serve the interests of residential and commercial developers over those of long-established (and often poorer) residents (Bennett et al., 2017). As in many North American cities, therefore, studying SUD in Chicago requires forthright acknowledgment of existing policies, practices, and associated processes.

In Chicago, for example, the process of gentrification tends to intersect with the built environment and racial boundaries in complex ways. Students quickly learn that so-called "green" development initiatives rarely include a social-equity component and instead often come packaged with higher rents, taxes, and displacement of existing residents (Curran & Hamilton, 2017). In the most theoretical of our core courses (SUD 401), therefore, students engage the city as a place to look at "sustainability" from a critical lens and explore why sustainability as a concept might be seen as problematic in certain neighborhoods. Such lessons become clearer as students explore the role of community voices in planning, whether it be brownfield remediation, affordable housing decisions, or the juxtaposition of neighborhood development and public transit. Students must ask: what does it mean to study SUD within a highly segregated city, where capital can quickly usurp community voice in neighborhood development decision-making? In addition, students work alongside faculty or university staff who are themselves engaged in research, training, and/or other forms of technical assistance with local governments, nonprofits, schools, and neighborhood associations. Consequently, students gain a sense of local politics and understand the reality of how cities change. In this way, 'sustainability' becomes a question applied through projects in the field, not a concept based on textbooks, marketing, or personal consumer decisions.

The emphasis on internships in the DePaul program sees students pursue a wide variety of interests with regard to the aspects of SUD. Typical placements range from positions at municipal, county, and federal government offices, to working for local nonprofits and neighborhood associations. To a lesser degree, students intern for private, for-profit urban architecture and design firms. The important role of these learning experiences is evident in the three following case studies, for which we have drawn upon students' written reflections. These cases exemplify how students have applied a critical lens on urban development through internship projects.

Transit-oriented Development

Transit-oriented development seeks to integrate housing, commercial, and retail spaces into a neighborhood that is walkable and in proximity to public transportation. One of the primary advantages of this type of development is that it promotes access to employment from housing that is intentionally developed to be affordable and amenable to walking, biking, and outdoor recreation. To explore the complexities of such an approach, one student completed an internship with a local nonprofit organization that seeks to integrate environmental sustainability, social equity, and technology with special emphasis on serving low-income communities. This student provided support for a campaign that explored transit solutions to increase accessibility for low-income residents to locations with higher levels of employment. Through research, the student found that many residents suffered from poor access to buses and trains, creating what the agency referred to as "transit deserts." Making matters worse, some of these areas also lacked employment opportunities, adding greatly to the plight of those living there.

Another intern designed and disseminated brochures describing these transit inequities to raise policy-maker awareness of these problems. In collaboration with the organization's staff, the student applied GIS software to spatially depict the juxtaposition of transit deserts and employment, population and density patterns, and proposed transit projects. This student recalled that the complexity of the work shaped the agenda of neighborhood organizations, explaining that "Working at a non-for-profit can be very rewarding, but also hectic and disorganized at times," because, the student continued, "each employee is constantly working on various projects that each individual feels very passionate about. This can lead to an interesting, yet frustrating, work environment."

Affordable Housing

Questions about housing, race, and economic equity in Chicago and most other United States cities are often divisive and laden with emotion. One of DePaul University's research centers, the Institute for Housing Studies (IHS), engages in research on housing in the region to inform planning and policy-making decisions. Affiliated with our university's Real Estate Center, IHS collects and analyzes data on affordability, homeownership, and neighborhood development, among other topics. The institute provides a resource for community organizations that seek to improve housing access. SUD students intern with the institute to apply and enhance research and data analysis skills learned through the curriculum.

One of the most salient examples of a student applying curricular skills at the institute involved a project entitled *Mapping Displacement Pressure in Chicago* (Institute for Housing Studies, 2017). Drawing on GIS techniques learned in the program, the student proposed this project idea during an IHS staff meeting as a way to offer an alternative perspective on neighborhood housing trends. Reflecting on how market forces and neighborhood goals are sometimes on a collision course, this study identifies areas of Chicago where residents are at risk of displacement due to rising property values. The students came to see why it was essential to support those "being disproportionately affected by these shifting market trends" in order to "ensure that sustainability doesn't become something just for a select few." "A city can be filled with parks, green buildings, etc.," the student observed, "but if there is a specific population of people that are priced out of living in these green buildings or living near these parks, then who is it sustainable for?"

Urban River Restoration and Redevelopment

Chicago has a well-publicized history with regard to managing its water resources, having grown up around the mouth of the Chicago River which, until city engineers reversed its direction in 1900, flowed eastwards into Lake Michigan. As a result of this undertaking and other ambitious waterway projects, the Chicago River was transformed into an industrial corridor in the late-1800s, with pollution and environmental devastation being notable downsides (Pacyga, 2009; Solzman, 2006). Today, rather than focusing on industrial uses, planners and developers seek to tap the river's potential for recreation and transportation, and to use adjacent land for residential and commercial purposes. As advocacy groups and planning agencies engage communities about the river's future, questions about the processes of restoration and redevelopment have ensued.

A student interning for a river restoration organization experienced this while coding and analyzing data provided by community members to support the waterway's restoration. Noting that responses from different neighborhoods tended to diverge, the organization was able to identify environmental justice issues that warrant policy-maker attention. The plan that the organization subsequently advocated not only addressed the restoration of the natural environment along the river but also the social risks, such as property tax increases from future development, which could negatively affect local residents. The intern came to see the conflict that sometimes exists between "environmental protection and economic growth and efficiency," and "environmental protection and social justice and income equality."

We contend that such experiences help students grasp the consequences of persistent income and racial divides which, in turn, foster perspectives on the necessity of rethinking how the concept of sustainability should be applied in urban planning, policy-making, and development. Although some SUD internships are housed in Chicago's vibrant downtown Loop district, they help students cultivate a more critical view of corporate-led development, particularly that which excludes or minimizes community voices and involvement.

Due to the newness of the DePaul program, the quality of student work has not yet been the subject of a large-scale faculty assessment. Although reflection papers are integral to the internship course, the program as a whole puts more emphasis on other deliverables, such as case-study evaluations, spatial analysis, and consulting-style reports with more narrowly focused learning goals. The capstone course (SUD 502) requires students to collaborate on a report about a particular neighborhood challenge that integrates concepts and skills learned throughout their studies. To date, we feel the program is doing a better job cultivating specialized learning (such as using GIS, measuring social inequities, and evaluating neighborhood issues) than in preparing students to conduct holistic assessments that broadly integrate theoretical and practical principles. Future assessments will reveal whether, in fact, this view is consistent with actual student performance.

BUILDING SUSTAINABLE URBAN DEVELOPMENT LEADERS

The degree to which the program creates career opportunities in SUD is reflected in formal academic review processes mandated by the university, as well as those required by higher education accrediting organizations. Engaging in these reviews requires answering questions about how or whether the program creates pathways for employment and leadership after graduation. Some such questions include: How do graduates apply the SUD degree professionally? What types of urban sustainability issues do they address? and How can program administrators measure the impact of the program, through its graduates, on cities?

Providing definitive answers to such questions is too ambitious a task for this chapter. Nevertheless, this final section begins this analysis by exploring the LinkedIn profiles of 30 of the 55 graduates of the program to date (Table 2). It seeks to examine the degree to which such profiles can paint a picture of the ways alumni are pursuing careers that potentially impact urban development.

The sample was not randomized and the content was not verified for accuracy, limiting its generalizability. Nevertheless, the results support the notion that SUD graduates are often employed in degree-relevant fields. Additionally, all the LinkedIn pages listed the SUD degree program as the most recent educational activity, suggesting that the majority of graduates are treating the program as a terminal degree. Few appeared to have attained senior management or executive management positions, though the majority was employed in areas closely related to urban development.

These findings give us optimism that our program is achieving, at some level, the goal of training professionals for careers in urban development. On a more

Table 2. Employment Experience Posted on LinkedIn by Graduates, Listed by Year of Graduation.

2015	Associate Planner (planning agency).
	Zoning and Permitting Specialist (telecommunications company).
	GIS Analyst (information services).
	Operations Manager (information technology and services).
	GIS Analyst (real estate).
	Market Analyst/Urban Planner (real estate consultants).
2016	Public Participation Planner (regional planning commission).
	Data Analyst (regional public transportation).
	Director of Zoning and Urban Development (city government).
	Program Associate, Property Preservation (affordable housing services).
	Maintenance Manager (rooftop and vertical farming services).
	Associate Analyst (utility company).
	Locations and Public Partnerships Coordinator (car sharing).
	Planner II (municipal government).
	Park Planner (county park district).
2017	Urban Design and Planning Consultant (urban planning consultants).
	Planner (civil engineering).
	Community Branding Manager (architecture and planning services).
	Mapping Logistics Planning Specialist (staffing and recruiting services).
	Project Coordinator (environmental services).
	Research Analyst (university research center).
	Regional Grants Officer (international nonprofit/funder).
	Marketing and Acquisitions (pharmaceuticals and medical supplies).
	Portfolio Administrator III (affordable housing services).
	Senior Research Associate (transportation policy and consulting).
2018	Logistics Associate (supply chain management).
	Development Associate (environmental planning and consulting).
	Coordinator of Applied Research (chamber of commerce).
	Project Manager (sustainability consulting company).
	Project Manager (municipal planning office).

cautionary note, though, only a small percentage appears to focus their work directly on issues of social inequality and inequity. Yet, it also appears that graduates are taking the technical knowledge of sustainability into leadership positions. Longer-term tracking of graduates, we hope, will allow for collecting more systematic evidence about the effectiveness of the SUD program.

CONCLUSION

Since its inauguration in 2013, DePaul's SUD graduate program has delivered an intentionally interdisciplinary curriculum, enabling students to draw upon a range of coursework from urban design to GIS to evaluate urban development through a critical lens. This lens is influenced by the social, economic, and political context of Chicago, and presents understandings of "sustainability" that go beyond narrow concentrations on the natural environment. SUD is about more than being "green." DePaul's SUD program is framed by a simple value –a sustainable city is one that is sustainable for *everyone*. This holistic approach encourages students to view urban environments as interconnected systems that function best when the entire community's voice is heard – and respected – in policy formation. Our curriculum seeks to strike a balance between skill-based, theoretical, and experiential learning, and this, we contend, produces graduates who are prepared to enter the workforce in mainstream sectors of urban development with a set of highly applicable skills and insights. Upon doing so, our hope is that these alums apply a critical lens in their professional roles and influence decision-making on urban development that advances environmental, racial, and economic equity in Chicago and beyond.

REFERENCES

Abu-Lughod, J. (1999). *New York, Chicago, Los Angeles: America's Global Cities*. Minneapolis, MN: University of Minnesota Press.

Bennett, L., Garner, R., & Hague, E. (Eds.). (2017). *Neoliberal Chicago*. Urbana, IL: University of Illinois Press.

Blewitt, J. (2004). Introduction. In J. Blewitt & C. Cullingford (Eds.), *The sustainability curriculum: The challenge for higher education* (pp. 1–9). Oxford: Earthscan.

Burgess, E. W. (1925). The growth of the city: An introduction to a research project. In R. E. Park, E. W. Burgess, & R. D. McKenzie (Eds.), *The city* (pp. 47–62). Chicago, IL: University of Chicago Press.

Curran, W., & Hamilton, T. (Eds.). (2017). *Just Green enough: Urban development and environmental gentrification*. New York, NY: Routledge.

Edwards, B. (2004). Sustainability and education in the built environment. In J. Blewitt & C. Cullingford (Eds.), *The sustainability curriculum: The challenge for higher education* (pp. 129–140). Oxford: Earthscan.

Fisher, C. (2015). *Urban Green: Nature, recreation, and the working class in industrial Chicago*. Chapel Hill, NC: University of North Carolina Press.

Institute for Housing Studies at DePaul University. (2017). Mapping displacement pressure in Chicago. Retrieved from https://www.housingstudies.org/releases/Mapping-Displacement-Pressure-in-Chicago/. Accessed on July 29, 2019.

Jones, P., Selby, D., & Sterling, S. (2010). More than the sum of their parts? Interdisciplinarity and sustainability. In P. Jones, D. Selby, & S. Sterling (Eds), *Sustainability education: Perspectives and practice across higher education* (pp. 17–37). Oxford: Earthscan.

Kevany, K., Huisingh, D., & Lozano García, F. J. (2007). Sustainability: New insights for education. *International Journal of Sustainability in Higher Education*, *8*(2), 103–106. https://doi.org/10.1108/ijshe.2007.24908baa.001

Klinenberg, E. (2015 [2003]). *Heat wave: A social autopsy of disaster in Chicago* (2nd ed.). Chicago, IL: University of Chicago Press.

Lorr, M. J. (2017). Urban sustainability and the 'Greening' of neoliberal Chicago. In L. Bennett, R. Gardner, & E. Hague (Eds.), *Neoliberal Chicago* (pp. 99–118). Urbana, IL: University of Illinois Press.

Pacyga, D. A. (2009). *Chicago: A biography*. Chicago, IL: University of Chicago Press.

Shabazz, R. (2015). *Spatializing blackness: Architectures of confinement and Black masculinity in Chicago*. Urbana, IL: University of Illinois Press.

Solzman, D. M. (2006). *The Chicago river: An illustrated history and guide to the river and its waterways* (2nd ed.). Chicago, IL: University of Chicago Press.

Sterling, S. (2009). Sustainable education. In D. Gray, L. Colucci-Gray, & E. Camino (Eds.), *Science, society and sustainability: Education and empowerment for an uncertain world* (pp. 105–118). New York, NY: Routledge.

United States Department of Homeland Security. (n.d.-a). STEM designated degree program list, effective May 10, 2016. Retrieved from https://studyinthestates.dhs.gov/eligible-cip-codes-for-the-stem-opt-extension and https://www.ice.gov/sites/default/files/documents/Document/2016/stem-list.pdf. Accessed on July 30, 2019.

United States Department of Homeland Security. (n.d.-b). Study in the states: Students: Determining STEM OPT extension eligibility. Retrieved from https://studyinthestates.dhs.gov/students-determining-stem-opt-extension-eligibility. Accessed on July 30, 2019.

United States National Center for Education Statistics/Integrated Postsecondary Education Data System. (2010a). Detail for CIP Code 45.1201. Retrieved from https://nces.ed.gov/ipeds/cipcode/cipdetail.aspx?y=55&cipid=88596. Accessed on July 30, 2019.

United States National Center for Education Statistics/Integrated Postsecondary Education Data System. (2010b). Detail for CIP Code 30.3301. Retrieved from https://nces.ed.gov/ipeds/cipcode/cipdetail.aspx?y=55&cipid=89257. Accessed on July 30, 2019.

Washington, S. H. (2005). *Packing them in: An archaeology of environmental racism in Chicago, 1865–1954*. Lanham, MD: Lexington Books.

Wille, L. (1991). *Forever open, clear, and free: The struggle for Chicago's lakefront* (2nd ed.). Chicago, IL: University of Chicago Press.

CHAPTER 7

AN INTERDISCIPLINARY TEACHING MODULE ON THE GLOBAL CLOTHING INDUSTRY: LESSONS FROM WORKING ACROSS FOUR DISCIPLINES AND TWO UNIVERSITIES

Niki Harré, Anis Azizi, Penny Brothers, Ties Coomber, Ellinor King, Andrea Michelle Mead, Sarah Saeckel, Manuel Vallée, Samantha Zi Lin Yeo and Yulun (Darren) Zhang

ABSTRACT

Since 2016, the authors have been teaching an interdisciplinary module on the global clothing industry to students enrolled in an introductory psychology course and a second year chemistry course at the University of Auckland in New Zealand. In 2016, the module also involved third year chemistry students, and in 2017, second year sociology students and graduate students in English literature from the University of Stuttgart in Germany took part. The module has the following features: (1) it focuses on a complex industry with ramifications for social and environmental sustainability, (2) it involves an issue of direct relevance to the students, (3) students teach those from another discipline as "subject experts," and (4) students are assessed on their learning

Integrating Sustainable Development into the Curriculum:
Innovations in Higher Education Teaching and Learning, Volume 18, 107–124

ISSN: 2055-3641/doi:10.1108/S2055-364120200000018025

within their home course. An evaluation of the 2018 iteration with psychology and chemistry students (N = 185) showed post-test decreases in participants' materialistic values, increases in knowledge and concern about the social and environmental impacts of the clothing industry and reported changes to clothes buying practices. The authors discuss the institutional barriers faced and provide five recommendations for other university teachers considering integrating an interdisciplinary sustainability module into existing courses.

Keywords: Interdisciplinary; clothing industry; sustainability; higher education; attitudes; knowledge

INTRODUCTION

Universities are central to global sustainability efforts. As learning institutions that offer high-level expertise across a range of disciplines, they are well positioned to develop students' understanding of the complex, multi-level issues that contemporary societies face, and how to address these. This understanding may, in turn, orient graduates toward building sustainable societies (Martin & Jucker, 2005).

In recognition of the role universities can play in forwarding sustainability, recent decades have seen a succession of international networks and agreements, including the 1990 *Talloires Declaration*, which currently has 505 signatories from universities in over 40 countries (University Leaders for a Sustainable Future, n.d.). The *Talloires Declaration* urges institutions of higher learning to create programs that would "produce expertise in environmental management, sustainable economic development, population, and related fields" and ensure "that all university graduates [be] environmentally literate and have the awareness and understanding to be ecologically responsible citizens" (University Leaders for a Sustainable Future, 1990). In 2009, The University of Auckland, where nine of us are located, signed the Universitas 21 *Statement on Sustainability*. This was an agreement with 20 partner universities to forward sustainability in all aspects of organizational life, including by becoming "living laboratories" and offering students education to "raise [their] sustainability consciousness" (Universitas 21, 2009). In 2017, our university also joined the *UN Sustainable Development Solutions Network*, a network designed to help implement the Sustainable Development Goals (Sustainable Development Solutions Network, n.d.).

Alongside these institutional agreements and aspirational goals, there has been a considerable increase in programs that offer environmental studies or a sustainability-related component. Such programs are now in place at most universities, and enrollments have been increasing (Adelman, 2004; Brint, Turk-Bicakc, Proctor, & Murphy, 2009; Frank, Robinson, & Olesen, 2011). These programs are often interdisciplinary and may offer entry level courses open to students from a range of programs (Caviglia-Harris & Hatley, 2004; Coops et al., 2015; Hoare et al., 2008). While this progress is positive and indicates growing *expertise* in environmental issues, universities often struggle to fully implement the ambitious educational vision behind institutional agreements such as those outlined (Stewart, 2017). In particular, they fall short on the aspiration that all, or at least the majority, of graduates gain the "sustainability consciousness" to become "ecologically responsible citizens." There are at least two significant barriers to this progress, both of which we experienced and attempted to overcome in the teaching module described in this chapter. One barrier is the relatively low priority universities that place on teaching innovation. This is in large part due to the institutional emphasis on research in most universities, and especially in "leading" universities (Harré, Grant, Locke, & Sturm, 2017). While research grants and outputs help build a university's position in international rankings and thus attract students, the emphasis on research also means limited effort is put into cutting-edge teaching (Haigh, 2005). Innovative sustainability teaching may therefore fail to get off the ground, simply due to the institution's attention being directed elsewhere.

Second, university teaching is almost always organized within the disciplinary silos that mirror the administrative structures of most universities (Coops et al., 2015; Haigh, 2005). Therefore, while sustainability educators consistently stress the need for an interdisciplinary approach to the environmental and social issues we face (Coops et al., 2015; Feng, 2012; Hoare et al., 2008; Howlett, Ferreira, & Blomfield, 2016; Stewart, 2017; Vincent & Mulkey, 2015), such approaches can be very difficult to establish. Issues include working with the different language of the disciplines concerned (Feng, 2012), varying traditions about the nature of assessment (Hoare et al., 2008), and a host of administrative barriers. It is notable, for example, that an interdisciplinary sustainability course at the University of British Columbia described by Coops et al. (2015) was only offered for one year. They cite a number of administrative reasons for this, a key one being that the course was not located within a single home faculty and so could not be credited toward students' degrees.

Nevertheless, we argue that it is worth persisting with interdisciplinary approaches to sustainability teaching and weaving these into existing programs. In this way, we give students who do not wish to specialize in an environmental discipline exposure to the complex issues we face and their possible solutions. This chapter describes an interdisciplinary teaching module on the global clothing industry that involved five courses from four different disciplines and spanned two universities. We first describe the history of the module and its various configurations, before outlining the current iteration. We then describe an evaluation of a recent iteration and the impact it had on the students who participated. We end this chapter by discussing reflections on, and recommendations for, integrating interdisciplinary sustainability teaching into existing university courses.

HISTORY OF THE TEACHING MODULE

The idea for a cross-course teaching module with a sustainability theme originally came from discussions within a sustainability network based in the Faculty of Science at The University of Auckland. The network includes academic and non-academic ("professional") staff from all nine disciplines located in the faculty. Meetings are held every six to eight weeks and members suggest projects, set up project teams, and report on progress. A teaching module was suggested by Andrea Mead, and initial conversations included psychologists (Andrea Mead and Niki Harré), a geographer (Joe Fagan), and a chemist (Penny Brothers). From the beginning, we decided that the module would be a case study focusing on a sustainability related issue and would involve a year 1 psychology course, years 2 and 3 green chemistry courses, and a year 2 geography course. After some weeks of discussion, we finally settled on the global clothing industry. There were three main reasons for this choice:

(1) The clothing industry is inherently complex. It involves both the environmental and social aspects of sustainability and all levels of the social system. We could each see multiple connections to our own discipline and were eager to learn about the disciplinary perspectives of each other.
(2) The clothing industry is directly relevant to the lives of *all* our students (unlike, e.g., cycling, which we also seriously considered). We hoped our students would be encouraged to experiment with their personal practices, recognize how these are shaped by global forces, and engage in meaningful discussions.
(3) A documentary film, *The True Cost,* had recently been released. This film shows how cotton is grown and the conditions of garment workers in countries such as India and Bangladesh, as well as exploring the advertising and consumption of fashion in rich industrialized nations such as the USA and the UK. As described by Sarah Saeckel (2016), the film has a vivid narrative and is emotionally compelling. Emotions help create student engagement (Walshe, 2017) and may reduce the psychological distance that facilitates social and environmentally problematic practices (Milfont, 2010; Van-Boven, McGraw, Kane, & Dale, 2010). This film is also information rich including, for example, reference to the impact of chromium-6, land-use and textile crops, and materialistic values. We saw the documentary as a way to provide the students from the different courses with a common foundation.

Unfortunately, having settled on the topic, the geography course that was originally going to participate (and was coordinated by someone who was not part of the project), pulled out. The coordinator explained that there had been unexpected changes to staffing that made it impractical to introduce a new module. A second geography course fell through as the topic didn't fit with the existing case study utilized in that course. We decided to go ahead with the psychology and chemistry courses and spent our next planning meetings working out how to run a shared module with courses in two different disciplines, at three different levels that had no shared timeslots.

Our first decision was to have all students watch *The True Cost* and have lectures on the clothing industry within their primary discipline. They would then give a presentation to students from the complimentary discipline and so be treated as "subject experts." We hoped, as discussed by Feng (2012), that our students would "make links between individual disciplines and generate cooperation between themselves as learners, becoming in the process a 'community of learning'" (p. 32). Our very large year 1 psychology course, with an enrollment of about 580 students, has 20 tutorial streams. This enabled students from the years 2 and 3 chemistry courses, which have about 50 and 25 of students, respectively, to be allocated a stream that fit their timetable and ensured they attended in pairs.

We also decided that students' learning would be assessed within their primary discipline. This allowed us to manage the different assessment styles of each discipline and the year level of the students. It also meant that we avoided the numerous administrative challenges that would result from attempting to mark shared assignments in a combined cohort of 650 students.

As of early 2019, the teaching module has been taught every semester since July 2016. The first year psychology course, which runs in both main teaching semesters (i.e., twice a year), has been involved in all five iterations and intends to continue with the module into the foreseeable future. The teaching module is also now a standard part of the year 2 chemistry course when it is taught in the second semester of each year. After the first iteration, the year 3 chemistry course was unable to participate as most of the students had experienced it in the previous year. For these students, a teaching module based on the same principles was introduced in collaboration with a year 3 science communication course. The theme was community engagement and the module featured a role-play in which the chemistry students were "scientists" presenting information about controversial topics to a community advisory group role-played by the science communication students.

In the first semester of 2017, a year 2 environmental sociology course with 65 students paired with the year 1 psychology course. Also in 2017, a graduate course in American/English literature from Stuttgart University joined for an online conference with The University of Auckland's teaching staff and a selection of students. The 21 students from Stuttgart University were studying *The True Cost* in the seminar called "Telling Stories for a More Sustainable and Fairer World." Before watching *The True Cost*, students read parts of Naomi Klein's *No Logo* and were asked to examine the omnipresent and emotionally effective advertising stories that keep people hooked on the status quo of overconsumption and neocolonial exploitation. In the online conference, 14 of the Stuttgart students presented an analysis of the film and showed how it subverts dominant storytelling and raises audiences' awareness for people and the planet. The Auckland students offered psychological and sociological perspectives on the clothing industry. We then had a wide ranging discussion about the issues raised. Sarah Saeckel, who taught the literature course was on a temporary contract, and so despite a lively exchange and the German students' enthusiasm about the topic, this component was not repeated. In semester 1, 2018, the psychology course did not have a partner course and so ran the module without the interdisciplinary component. Given the number of iterations, for simplicity, we now briefly describe the current interdisciplinary iteration which involves first year psychology students and second year green chemistry students.

DESCRIPTION OF THE CURRENT INTERDISCIPLINARY ITERATION

The module begins with students from both the psychology and chemistry course completing an online pre-test survey (more detail is given in the Evaluation section to follow). Students then watch *The True Cost* in class or via a streaming link provided. The psychology students are asked to note down both the psychological dimensions mentioned in the film and the questions they have for chemistry students. Next, students receive a set of lectures in their enrolled course and from their disciplinary perspective. Psychology students learn about the role of *materialistic values* and *psychological distance* in promoting fast fashion in Western nations. Chemistry students learn about the energy, water, and pesticide burdens for growing cotton, as well as green chemistry solutions for sustainable fiber and clothing production.

Pairs of chemistry students are assigned to 1 of the 20 psychology tutorial streams. At the tutorials, the chemistry students give a single presentation and the psychology students give a series of presentations in groups of five or six. Each presentation must cover two problems and two solutions associated with the clothing industry. Questions and discussions between the students follow. Students are assessed on what they have learnt from the students from the other course through short answer questions in the exam (psychology) or through a written reflection (chemistry). Psychology students also complete a written assignment that involves coding the responses to an open-ended question on the pre-test survey that asks students what clothing means to them. Finally, students are invited to complete a post-test survey (details to follow).

EVALUATION OF THE 2018 PSYCHOLOGY/CHEMISTRY ITERATION

Although we did pre-/post-surveys every time the module was taught, we include here only data from semester 2, 2018, as it involved the most comprehensive measures and focused on the latest interdisciplinary iteration (as described in the previous section). The purpose of the evaluation was to understand students' response to the teaching module. In particular, we had four key questions:

(1) Did students' knowledge of the social and environmental effects of the clothing industry increase after the module?
(2) Did students' attitudes change before and after the module? In particular, we were interested in the following:
   (a) The extent to which they endorsed materialistic values (or materialism). Materialism is defined by Richins and Dawson (1992) as a value orientation that considers the acquisition and ownership of material goods as an essential means to achieve life goals. Clothing is one of the key material means by which young people manage their identities (Joy, Sherry, Venkatesh, Wang, & Chan, 2012) and the problems with

materialistic values are covered at some length in *The True Cost*, the psychology teaching, and in the psychology students' presentations to the chemistry students. Therefore, we thought that they may become less materialistic after the module.

(b) Their general concern for environmental and social issues.

(c) Their clothing specific environmental and social concern.

(3) Did students' clothes-buying habits change after the module?

(4) How did students respond to each element of the module and the module as a whole?

Given that students were provided with knowledge about the clothing industry in lectures, via *The True Cost* and by students from the other discipline, we expected to see increases in knowledge. For the reasons given in (2) above, we also expected decreases in materialism, increases in both general and clothing-specific social and environmental concern, and possibly changes to clothes-buying habits.

METHOD

Participants and Procedure

In total, 352 students completed the pre-test and 385 students completed the post-test. For the pre-/post-measures, we used only students that could be matched. Due to a requirement from our ethics committee that anonymity be preserved, students created their own codes, and while we suggested a formula to them and asked them to photograph or record their code, we could only match 185 students in total. For the items that assessed students' response to the teaching module (related to research question (4) given above), we used the data from all students who completed the post-test.

As can be seen in Table 1, 185 students from psychology ($n = 179$) and chemistry ($n = 6$) were included in the pre-/post-data analysis. Of the 185 matched participants, 72% identified as female, and 95% were aged between 18 and 25 years. Most were in their first year at university and identified as "New Zealanders" (37%), followed by "Asian" (23%). Approximately 80% had $100 or less of disposable income per week after essential living expenses. These demographics were similar in the full post-test group ($n = 385$). However, one difference was that the proportion of participants who identified as Asian was larger in the post-test group (32%), and only 30% identified as New Zealanders.

All the surveys were completed online. For the pre-test, an invitation and link was sent to students via email two weeks before the teaching began, with a reminder one week later. For the post-test, chemistry students were emailed the invitation to participate and link at the end of the module. The psychology students were invited to do most of the post-test in class before the student presentations. A class setting was chosen to boost the post-test response rate, which was significantly lower than the pre-test response rate in previous iterations of the module. Unfortunately, it was not possible to schedule the post-test later in

Table 1. Demographic Characteristics of Participants in the Evaluation.

Demographics	Pre-/post-matched Participants ($N = 185$)		All Post-test Participants ($N = 385$)	
	%	*n*	%	*n*
Gender				
Male	26.0	48	28.8	111
Female	72.4	134	69.6	268
Gender diverse	1.1	2	0.8	3
Prefer not to answer	0.5	1	0.3	1
[Missing]	–	–	[0.5]	[2]
Age				
18–25	95.2	176	95.1	366
26–39	4.3	8	3.9	15
40–48	0.5	1	0.5	2
[Missing]	–	–	[0.5]	[2]
Year at university				
First year	79.5	147	77.9	300
Second year	14.6	27	14.0	54
Third year or more	5.9	11	7.3	28
[Missing]	–	–	[0.8]	[3]
Course				
Year 1 psychology	96.8	179	97.1	374
Year 2 chemistry	3.2	6	2.9	11
Ethnicity				
New Zealander	37.3	69	30.1	116
Asian	23.2	43	32.5	125
Other/mixed	22.2	41	20.5	79
Pacific Peoples	8.1	15	4.7	18
European	5.9	11	7.0	27
Middle Eastern/Latin American/African	2.2	4	3.6	14
New Zealand Māori	1.1	2	1.6	6
Weekly disposable income				
$25 or less	21.6	40	18.4	71
$26–$50	27.6	51	26.5	102
$51–$100	31.4	58	31.2	120
$101–$200	14.0	26	16.1	62
$201 or more	5.4	10	7.3	28
[Missing]	–	–	[0.5]	[2]

Note: The ethnicity identities match the ones used to construct the NZ Census, except for "New Zealander" which was an identity we created to reflect how many students described themselves. Participants could tick as many ethnic identities as they wished.

the course. However, psychology students were emailed an invitation and link to the questions specifically on the student presentations at the end of the module.

Measures

Students were first given an information sheet approved by The University of Auckland Human Participants' Ethics Committee and asked to check if they agreed to their responses being used for research purposes. Four sets of measures were used to compare the pre-/post-responses of participants.

Knowledge of the environmental and social impacts of the clothing industry was measured with the following three open-ended questions: (1) What do you know about the environmental effects caused by the growing of crops used for clothing, such as cotton? (2) What do you know about the environmental effects caused by the manufacturing of clothes? (3) What do you know about the conditions of clothing workers? Given that participants often provided information relevant to one of the questions in the response box for another question, we combined each participant's response to all three questions for coding purposes. Four categories of knowledge, relevant to the course content, were created. They were knowledge of: (1) resource depletion, (2) environmental damage, (3) physical conditions of clothing workers, and (4) social conditions of clothing workers. A point was awarded for every separate fact or example, for a maximum of three points per category. This created a maximum score of three for each category of knowledge, a maximum score of six for knowledge of social impacts and knowledge of environmental impacts, and a maximum of 12 for total knowledge. One researcher coded all responses, and another researcher used the same coding scheme to code one third of responses. The intercoder reliabilities were high, with an agreement rate of 99.6% for the pre-test, and 98% for the post-test. Disagreements were discussed until agreement was reached on a suitable code.

Materialistic values were measured using the materialism scale developed by Richins and Dawson (1992). The scale has 18 items and 3 subscales that measure how much participants consider possessions define *success,* how central materialism is to a participants' identity (*centrality*), and how much *happiness* material goods provide. In the current study, the scale showed good reliability, with a Cronbach's alpha of 0.86 in the pre-test and 0.87 in the post-test. The Cronbach's alphas of the three materialism subscales were: *success* 0.72 pre-test and 0.75 post-test, *centrality* 0.75 pre-test and 0.71 post-test, and *happiness* 0.76 pre-test and 0.82 post-test.

General environmental concern and social concern were each measured with a single item using a sliding scale of 0–100. Participants were asked, "How much do you care about the following: 1) Environmental issues e.g., climate change, water shortages, pollution, etc. and 2) Social justice issues e.g., fair trade, employment conditions, health inequality, etc."

Clothing-specific environmental concern and social concern were measured using three five-point Likert scale items, two for environmental concerns and one for social concerns. Participants were asked:

> Before making a clothing purchase do you ever consider the following: 1) The environmental effects caused by the growing of crops used for clothing, 2) The environmental effects caused by manufacturing clothing, and 3) The conditions of the workers who made the item.

The responses for the two environmental items were averaged to create a single score for specific environmental concern.

Clothes-buying behavior. Participants were asked six questions on their clothes-buying habits. However, given the limited time between the teaching and the post-test, only the two questions on intentions were analyzed for the evaluation. Response options are given after the question.

- Over the next six months, how often do you think you'll buy clothes? (1) Weekly, (2) Fortnightly, (3) Monthly, (4) Twice, and (5) Once or not at all.
- Over the next six months, when shopping for clothes, how many items do you think you will buy? (1) 1–2 items, (2) 3–4 items, and (3) 5+ items.

Item 1 was reverse scored, so that a high score represented more clothes purchased, consistent with item 2.

In the post-test only, students were asked how valuable or helpful they found each component of the teaching module: (1) *The True Cost*, (2) the Lectures, (3) the assignment, (4) giving a presentation, (5) hearing presentations, and (6) the Q & A with students from the other course. For questions (1)–(3), the response options ranged from 1 (not helpful at all) to 5 (very helpful). For questions (4)–(6), the response options ranged from 1 (not valuable) to 5 (very valuable).

An open-ended question at the end of the post-test was asked to students: "If the teaching did affect your attitudes or clothing buying habits can you please explain how?" Thematic analysis was used to identify common patterns in the data. Responses could be coded into more than one theme.

Results

Table 2 shows the mean responses to the pre- and post-test knowledge and attitude measures for the 185 participants whose responses could be matched.

In regard to *knowledge of environmental impacts*, paired t-tests found significant mean increases in knowledge of environmental damage associated with the clothing industry (mean difference $= -0.93$, $t[184] = -10.52$, $p < 0.001$) and in the summed total of depletion and damage knowledge (mean difference $= -1.02$, $t[184] = -8.18$, $p < 0.001$). There was no significant change in the knowledge of resource depletion after the module. In regard to *knowledge of social impacts*, there were significant mean increases in the knowledge of problematic physical conditions for clothing workers (mean difference $= -0.61$, $t[184] = -7.15$, $p < 0.001$) and in the summed total knowledge of problematic social and physical conditions for workers (mean difference $= -0.81$, $t[184] = -5.47$, $p < 0.001$). There was only a borderline significant mean difference in the knowledge of social problems for clothing workers (mean difference $= -0.20$, $t[184] = -1.90$, $p = 0.06$). Finally, there was a significant increase in overall knowledge (mean difference $= -1.82$, $t[184] = -8.36$, $p < 0.001$).

The results of a paired t-test found a significant average decrease in *materialistic values* after the teaching module (mean difference $= 0.06$, $t[184] = 2.34$, $p = 0.02$). Paired t-tests for the three subscales suggest that the significant mean difference was underpinned by a significant mean decrease in *centrality* (mean difference $= 0.10$, $t[184] = 2.66$, $p = 0.01$). There was no significant change in the *success* subscale, and the happiness subscale showed a borderline significant mean decrease of 0.08 ($t[184] = 1.92$), $p = 0.06$.

Inconsistent with expectations, the results of paired t-tests found significant average decreases in *general environmental concern* (mean difference $= 2.94$, $t[184] = 2.60$, $p = 0.01$) and in *general social concern* (mean difference $= 2.81$, $t[184] = 2.28$,

Table 2. Students' Knowledge and Attitudes Before and After the Teaching Unit.

	Pre-test ($N = 185$)			Post-test ($N = 185$)		
Scale and Subscale	*M*	SD	SE	*M*	SD	SE
Knowledge of environmental and social impacts of clothing industry (Scale 0–12)	3.05	2.60	0.19	4.87	2.67	0.20
Environmental impacts (Scale 0–6)	1.21	1.49	0.11	2.23	1.57	0.12
Resource depletion	0.25	0.61	0.05	0.34	0.91	0.07
Environmental damage	0.96	1.13	0.08	1.89	1.13	0.08
Social impacts (Scale 0–6)	1.84	1.54	0.11	2.64	1.70	0.12
Physical conditions	0.54	0.77	0.06	1.14	0.98	0.07
Social conditions	1.30	1.09	0.08	1.50	1.08	0.08
Materialist values (Scale 1–5)	2.88	0.58	0.04	2.82	0.58	0.04
Success	2.66	0.69	0.05	2.67	0.67	0.05
Centrality	3.00	0.66	0.05	2.90	0.61	0.04
Happiness	2.96	0.78	0.06	2.88	0.85	0.06
General environmental and social concern (Scale 0–100)						
General environmental	74.34	19.48	1.43	71.40	20.86	1.53
General social	76.40	19.85	1.46	73.60	21.96	1.61
Specific environmental and social concern (Scale 1–5)						
Specific environmental	2.06	1.10	0.08	2.51	1.12	0.08
Specific social	2.97	1.25	0.09	3.29	1.16	0.09

Note: *M*, SD, and SE represent mean, standard deviation, and standard error, respectively. For all four environmental effects and social impact knowledge subscales, scores range from 0 to 3.

$p = 0.02$). However, consistent with the hypotheses, paired *t*-tests found significant increases in both *specific environmental concern* (mean difference = –0.45, $t[184] = -5.81$, $p < 0.001$) and *specific social concern* (mean difference = –0.32, $t[184] = -3.75$, $p < 0.001$).

Paired *t*-tests were conducted on the two measures of *clothes-buying behavior*. There was a significant pre-/post-change in the expected purchase quantity over the next six months (mean difference = 0.15, $t[184] = 2.24$, $p = 0.03$), but no change in the number of items they expected to buy on each occasion.

Table 3 shows the themes constructed from the students' responses to the question, "If the teaching did affect your attitudes or clothing buying habits can you please explain how?"

The most common response was increased awareness of the issues, found in half of the responses. For example, one respondent wrote, "Raised my awareness and taught me the actual implications of clothes buying. Brought it to the forefront of my mind." Just under 10% of participants used emotional terms to describe how they were affected, most commonly guilt. For example, "I used to get a rush at the thought of paying little money for items I wanted. Now I

Table 3. Students' Responses to the Open-Ended Question on How the Teaching Affected Their Attitudes or Clothes Buying Habits.

Theme	# (%)	Description
Increased awareness of the issues	138 (50.2)	Indicated greater awareness or knowledge of the issues connected to clothing production
Change in clothes-buying behavior	73 (26.7)	Indicated a clear change in buying behavior, for example, changes in where they shop, buying frequency, and/or type of clothing items purchased
Affected the clothes buying process	60 (22.0)	Affected their thought process before making a clothing purchase, with changed priorities and new factors to consider
Desire to change behavior	32 (11.7)	Indicated that they were "wanting to," "trying to," and/or would "like to" change their behavior
Emotional reaction	27 (9.9)	An emotional response, for example, that they felt guilt, fear, or sadness in relation to clothing production
No change	26 (9.5)	Indicated no impact on their behavior, and they would still continue with their clothing-buying habits
Supported existing behavior	21 (7.7)	Indicated that their current attitudes and clothing-buying habits already reflect awareness of the issues and were reinforced or supported by the content of the teaching
Reduced psychological distance	17 (6.2)	Made connections between their own clothing practices and previously distant issues with clothing production

Note: 112 provided no response, resulting in a total of 273 respondents, 102 of the respondents indicated more than one theme.

no longer have the same feeling, I feel ashamed and guilty." A small percentage (6.2%) indicated the module had decreased the psychological distance between them and the issues connected to clothing production, as shown in the following example:

> It has really shown me how big of an influence we have on others' lives, which is something we don't think about. Now, when I consider buying a piece of clothing that is on sale or is cheap, I remember what the working conditions for those people are how much they suffer. I also realise how precious clothing can be, and it is not something that is so easily disposable.

Notably many respondents (22%) referred to thoughts they had when considering clothing purchases, indicating that these changes in attitudes and emotions may have influenced the clothes-buying process. This can be seen in the previous example and here,

> Now when I think about buying new clothes I take into account who I am buying it from, is it environmentally and worker friendly and do I really need it or am I being unsustainable or merely conforming to social norms in society?

Others (11.7%) indicated a desire to change their behavior, for example,

> I am more aware of the issue and would like to see what I could do to stop it. Hopefully I will make more environmental/ethical clothing choices in the future and participate in further social action.

Nearly 27% specifically mentioned an actual change in their clothes buying. Examples included changes to where they choose to shop for clothes, reduction in clothing consumption, and buying environmentally friendly or ethically made products. Here is one,

> I don't shop from certain stores anymore. I am also more aware of what I buy and how often I choose to shop for clothes. The frequency that I buy clothes has decreased a lot as my consumption before was high.

A further (7.7%) commented that there is no need to change their clothes-buying habits as they are consistent with the teaching provided, for example,

> The teaching helped my understanding of the process behind my clothes. It changed my perspective on how people buy clothes however I don't really buy a lot of clothes to begin with so the change in perspective has not changed my habits.

Despite sometimes indicating the teaching had increased their awareness, 9.5% of participants commented that their behavior was unlikely to change. Often this was because "ethical" clothes were considered too expensive.

> It made me think more about the origins of the clothes I buy but I doubt my clothing buying habits will change a lot as clothing from more environmentally friendly and ethical sources is much more expensive and I don't really have the budget for it right now.

Others that commented they would not change their clothes-buying practices spelt out the complexities of making change in the social context of their lives. Here is one:

> It has changed my attitude a little but it has not changed my habits as a lot of my habits for this are not based off my attitude but the necessities in order to maintain my job. I work in a chain retail store and we are required to buy clothing almost fortnightly to stay with the brands fashion output. I never bought many items of clothing so frequently until I started working here however I wish I didn't have to. It's a tricky balance between ethics and attitude towards clothing, but also survival. Being a student and having a part-time job really is vital to my day-to-day life.

For another participant, the issue was seen as political, making their own behavior irrelevant,

> I find myself hesitating on wanting to buy things because of the chemicals however a lot of the time I forget because I'm thinking more about how me boycotting the clothing industry isn't going to change anything so I just go ahead and spend my money anyway.

Table 4 shows the post-test participants' rankings of how helpful or valuable they found each aspect of the teaching. As can be seen, *The True Cost* was considered extremely helpful, followed by the lectures and then writing the assignment. The three items that measured the exchange with students from the other course had somewhat lower rankings.

Discussion

Our first research question was whether students' knowledge of the social and environmental impacts of the clothing industry increased after the teaching module. Knowledge increases were indeed found, particularly with regard to the environmental damage caused by the industry and the physical conditions for clothing workers. It is notable that for the psychology students, this information was not covered in the lectures and so will have come from other parts of the module.

Table 4. How Helpful Students Considered Each Component of the Teaching Unit.

Questions	*n*	*M*	SD
1. How helpful was *The True Cost* documentary?	184	4.49	0.82
2. How helpful were the lectures on clothing?	184	3.88	1.01
3. How helpful was the writing assignment?	184	3.57	1.06
4. How valuable to your learning was giving a presentation to students from the other course?	32	2.84	1.22
5. How valuable was hearing students from the other course present?	32	3.45	1.06
6. How valuable was having a Q & A with students from the other course?	32	3.03	1.33

Note. $N = 185$. For questions (1)–(3), scores range from 1 (not helpful at all) to 5 (very helpful). For questions (4)–(6), scores range from 1 (not valuable) to 5 (very valuable). The lower *n* for questions (4)–(6) reflects that the psychology students completed these questions separately. This is explained in the Method section of the evaluation.

In regard to attitudes, we found a small decrease in materialistic values after the module, particularly with regard to how central materialism was to participants' identity. We also found increases in clothing-specific social and environmental concern. However, we found slight *decreases* in general social and environmental concern. One explanation for these seemingly contradictory findings may be that clothes are a common means by which young people express their identity (Joy et al., 2012), and both the psychology teaching and *The True Cost* emphasized this. Hence, the teaching directly targeted materialism. *The True Cost* and the chemistry teaching also emphasized the social and environmental issues with the clothing industry. However, none of the teaching covered social and environmental issues at a general level. This may be why we found positive changes on materialism and clothing-specific concern, but not on general concern. The decreases we found in general concern, while small, may have implications for sustainability teaching as will be discussed in the next section.

Participants indicated an expected decrease in clothing purchase frequency, but not an expected decrease in the number of items purchased. Data from the open-ended question in the post-test provide further insight into how the teaching influenced students' clothes buying. Many reported being much more aware of the social and environmental costs of clothing, with some noting changed emotions, particularly a feeling of guilt, when considering a purchase. Over a quarter mentioned changes to clothes-buying habits. These changes went well beyond what was captured in our quantitative measures and included reconsidering where they shop and the ethical issues associated with particular garments.

While the Stuttgart students from an earlier iteration of the module did not complete a formal evaluation, they also informed their teacher, Sarah Saeckel, that they had been affected by the teaching. Some said they had developed a conscience about buying clothes and particularly fast fashion. One student wrote and performed a song about the clothing industry with the refrain, "We can do better than this," which is a direct quotation from *The True Cost*. Two others were considering setting up an Instagram account to promote eco-clothing options.

The evaluation described here was with only one iteration of the module, and it had limitations. First, only 185 participants could be matched on the pre- and post-tests, and only six of these were chemistry students. This is partly due to the much lower number of chemistry students overall, but nevertheless means that the experience of students in this course was not effectively captured. Second, there was a gender imbalance, with 72% of the students identifying as female. Third, the psychology students completed most of the post-test prior to the exchange with chemistry students. This means that the evaluation does not adequately reflect the learning from that component. Nevertheless, the psychology students had constructed questions for the chemistry students while watching *The True Cost* and begun work on their presentations to these students. We will pick up on the participants' ratings for the different module components in the general discussion to follow.

GENERAL DISCUSSION

We have outlined the history, description, and evaluation of a teaching module on the clothing industry that has spanned two universities and three disciplines and is now a standard part of a large introductory psychology class and a year 2 green chemistry class. The module appears to have prompted many students to think deeply about a complex, real world issue with implications for both social wellbeing and the natural environment. Furthermore, many students reported that the issues covered in the module were in their minds when buying, or thinking about buying, clothes. This indicates a reduction in the psychological distance between them and the production of their clothes (Milfont, 2010; Van-Boven et al., 2010). It also shows that the teaching leaked into students' everyday practices, encouraging critical engagement with social systems and the transformative, values-based changes. Many scholars have suggested that it is at the heart of university-based education for sustainability (Cortese, 2003; Feng, 2012; Haigh, 2005; Scott, 2009).

For us as teachers, the interdisciplinary aspect to this module was novel and stimulating. It required us to think creatively and cooperatively about what to teach and how to teach it, and it was fascinating to learn what the other disciplines had to offer. We realized too, that although much of the teaching was within the students' primary discipline, the whole module was infused with its interdisciplinary underpinning. For example, we taught material that we felt our students could teach to other students, and the psychology students analyzed qualitative data from the pre-test in their accompanying assignment that included the responses of chemistry or sociology students.

As we were committed to designing an interdisciplinary sustainability teaching module, we dealt with the numerous difficulties associated with teaching across disciplines, year levels, countries, and time zones as they arose. Many initial difficulties became solutions. For example, having a ratio of 1:15 chemistry to psychology students, which at first seemed to mitigate against peer-to-peer teaching, meant we could include pairs of chemistry students in the psychology tutorials without needing larger rooms (which would have been impossible to arrange given space limitations). The 12-hour time difference between Auckland and Stuttgart, which prevented us from holding the online conference in normal working hours in either location, also meant we could arrange a morning/evening session that was an easier fit with all our schedules.

Despite our best efforts however, we were still beset with some institutional barriers. Our psychology/sociology pairing lasted only one semester and was, perhaps counterintuitively, more awkward than the psychology/chemistry pairing. As Feng (2012) has argued, one of the tensions experienced in interdisciplinary collaborations is the different terminology used, and the struggle students experience to integrate a new language. When the psychology and sociology courses were paired this struggle was particularly apparent as the two disciplines often use different terminology to explain overlapping concepts. For example, the psychology students had learnt about "materialistic values" and the sociology students about "consumerism." The language of the chemistry students (who talked of elements and molecules; farming and technical processes), however, was so different to that of the psychology students, that it was relatively easy for each to appreciate and learn the others' concepts.

The sociology coordinator, Manuel Vallée, was also invited into the collaboration after the first iteration, meaning that he had less opportunity to shape the content. While he and many of his students enjoyed the additional dimension provided by the interdisciplinary exchanges, he did not renew the collaboration the following year because the assessments for the module displaced assessments that were better aligned with his course objectives. Fit was also an issue for one of the geography courses we approached, and the difficulties posed by a sudden change in staffing for another. Finally, Sarah Saeckel's temporary contract meant the Auckland/Stuttgart pairing could not be repeated. While each of these incidents is, in itself, unique, when put together they also show the difficulty of unpicking the "barriers of specialization" (Haigh, 2005, p. 38) university structures create. On the one hand, teachers are encouraged to have a strong, individual, vision for their teaching, while on the other hand, many courses are taught by staff who come and go. Both these situations create obstacles to teaching innovation across disciplines.

While our module as a whole was successful, students rated the peer-to-peer presentations as less useful to their learning than *The True Cost* documentary or the within discipline teaching. This was disappointing and may have been due to the power and comprehensiveness of *The True Cost*. By the time students heard from each other, they had seen the documentary and already been exposed to the issues raised through a vivid, emotionally compelling medium. To make the presentation component more compelling, students could present material that would be new to their peers. It might also help to create an assessment that weighted this component more heavily than in our courses (where it comprised less than 5% of the course marks).

Another limitation with our approach is that we did not increase students' general social and environmental concern, in fact this may have *decreased* as a result of the teaching. We wonder if, by introducing them to an issue most had not considered in depth before, we took their attention away from issues such as climate change which are currently culturally salient. We could remind students that there are many other serious issues too, but we feel a better response would be for more courses to include reference to social and environmental issues and thus ensure these too are made salient. Also, in and of itself, "concern" is perhaps not what we are aiming for here, but instead for our students' to realize their role in the structures that cause environmental and social problems and consider alternatives.

We have a number of suggestions for other teachers who would like to include an interdisciplinary component on a sustainability issue in existing courses. First, do it! While the logistics can initially seem insurmountable, as long as courses are scheduled in the same semester, these can probably be overcome. Second, a strong, shared stimulus provides an excellent anchor point for the teaching. In our case, a documentary *The True Cost* ensured that all the students had a common experience. A recent news event could also be used or a case study of the local area (see Steiner & Posch, 2006). Third, it is important to assess the students' learning, but this can be done within their home course. This greatly simplifies the challenges of working across disciplines and year levels. Fourth, it may be easier to work across very different disciplines, than across similar disciplines. The language of similar disciplines tends to overlap and thus compete, the language of different disciplines does not. Fifth, we left many students with an uncomfortable sense of disorientation in regard to clothes buying. While some disorientation is essential to transformative learning (Mezirow, 1978), in future we intend to provide links to organizations that promote fair and eco-friendly clothing production. We recommend that teachers who create similar modules also consider if and how to direct their students toward appropriate action options.

To conclude, universities can play a vital role in advancing sustainability knowledge, awareness, and action, and we argue that they should take this role seriously. One way to do so is for teaching staff to work together on interdisciplinary partnerships. Cross-course teaching modules, such as the one discussed here, not only expose students to useful information from outside their primary discipline, but also give them insight into the complexity of sustainability issues and the need for multi-faceted solutions.

REFERENCES

Adelman, C. (2004). *The empirical curriculum: Changes in postsecondary course-taking, 1972–2000*. Washington, DC: US Department of Education.

Brint, S., Turk-Bicakc, L., Proctor, K., & Murphy, S. P. (2009). Expanding the social frame of knowledge: Interdisciplinary degree-granting fields in American colleges and universities, 1975–2000. *The Review of Higher Education, 32*(2), 155–183.

Caviglia-Harris, J. L., & Hatley, J. (2004). Interdisciplinary teaching: Analyzing consensus and conflict in environmental studies. *International Journal of Sustainability in Higher Education, 5*(4), 395–403.

Coops, N. C., Marcus, J., Construt, I., Frank, E., Kellett, R., Mazzi, E., … Sipos, Y. (2015). How an entry-level, interdisciplinary sustainability course revealed the benefits and challenges of a university-wide initiative for sustainability education. *International Journal of Sustainability in Higher Education, 16*(5), 729–747. doi:10.1108/IJSHE-04-2014-0059

Cortese, A. D. (2003). The critical role of higher education in creating a sustainable future. *Planning for Higher Education*, *31*(3), 15–22.

Feng, L. (2012). Teacher and student responses to interdisciplinary aspects of sustainability education: What do we really know? *Environmental Education Research*, *18*(1), 31–43. doi:10.1080/13504622.2011.574209

Frank, D. J., Robinson, K. J., & Olesen, J. (2011). Expansion of environmental education in universities. *Comparative Education Review*, *55*(4), 546–573.

Haigh, M. (2005). Greening the university curriculum: Appraising an international movement. *Journal of Geography in Higher Education*, *29*(1), 31–48.

Harré, N., Grant, B., Locke, K., & Sturm, S. (2017). The university as an infinite game: Revitalising activism in the academy. *Australian Universities' Review*, *59*(2), 5–13.

Hoare, A., Cornell, S., Bertram, C., Gallagher, K., Heslop, S., Lieven, N., … Willmore, C. (2008). Teaching against the grain: Multi-disciplinary teamwork effectively delivers a successful undergraduate unit in sustainable development. *Environmental Education Research*, *14*(4), 469–481. doi:10.1080/13504620802278811

Howlett, C., Ferreira, J.-A., & Blomfield, J. (2016). Teaching sustainable development in higher education: Building critical, reflective thinkers through an interdisciplinary approach. *International Journal of Sustainability in Higher Education*, *17*(3), 305–321.

Joy, A., Sherry, J. F., Venkatesh, A., Wang, J., & Chan, R. (2012). Fast fashion, sustainability, and the ethical appeal of luxury brands. *Fashion Theory*, *16*(3), 273–295.

Martin, S., & Jucker, R. (2005). Educating Earth-literate leaders. *Journal of Geography in Higher Education*, *29*(1), 19–29.

Mezirow, J. (1978). Perspective transformation. *Adult Education*, *28*(2), 100–110.

Milfont, T. L. (2010). Global warming, climate change and human psychology. In V. Corral-Verdugo, C. H. García-Cadena, & M. Frías-Arment (Eds.), *Psychological approaches to sustainability: Current trends in theory, research, and practice* (pp. 19–42). New York, NY: Nova Science.

Richins, M. L., & Dawson, S. (1992). A consumer values orientation for materialism and its measurement: Scale development and validation. *Journal of Consumer Research*, *19*(3), 303–316.

Saeckel, S. (2016). 'Dirty clothes', their 'Greenwashing' and a call for change in Andrew Morgan's film The True Cost (2015). *Context*, *32*(Winter), 54–68.

Scott, W. (2009). Environmental education research: 30 years on from Tbilisi. *Environmental Education Research*, *15*(2), 155–164. doi:10.1080/13504620902814804

Steiner, G., & Posch, A. (2006). Higher education for sustainability by means of transdisciplinary case studies: An innovative approach for solving complex, real-world problems. *Journal of Cleaner Production*, *14*, 877–890.

Stewart, G. (2017). Securing knowledge of sustainability in 21st century higher education. *Solutions*, *8*(1), 44–53.

Sustainable Development Solutions Network. (n.d.). Sustainable-development-solutions-network. Retrieved from http://unsdsn.org/. Accessed on February 27, 2019.

Universitas 21. (2009). Statement on sustainability. Retrieved from https://cdn.auckland.ac.nz/assets/auckland/about-us/the-university/sustainability-and-environment/universitas-21-statement-on-sustainability.pdf. Accessed on February 27, 2019.

University Leaders for a Sustainable Future. (1990). The Talloires Declaration 10 point plan. Retrieved from http://ulsf.org/wp-content/uploads/2015/06/TD.pdf. Accessed on February 27, 2019.

University Leaders for a Sustainable Future. (n.d.). Talloires Declaration. Retrieved from http://ulsf.org/talloires-declaration/. Accessed on February 27, 2019.

Van-Boven, L., McGraw, A. P., Kane, J., & Dale, J. (2010). Feeling close: Emotional intensity reduces perceived psychological distance. *Journal of Personality and Social Psychology*, *98*(6), 872–885.

Vincent, S., & Mulkey, S. (2015). Transforming US higher education to support sustainability science for a resilient future: The influence of institutional administrative organization. *Environment, Development and Sustainability*, *17*, 341–363.

CHAPTER 8

INTEGRATING SUSTAINABILITY INTO A FRESHMAN-ENGINEERING COURSE THROUGH AN INSTITUTE-LEVEL INITIATIVE: A TEACHING–LEARNING MODEL WITH AUTHENTIC ACTIVITY AND CONTEXT

Raghu Pucha, Kata Dosa, Sunni Newton,
Meltem Alemdar, Ruthie Yow and Jennifer Hirsch

ABSTRACT

In January 2016, Georgia Tech launched a campus-wide academic initiative ("Center for Serve-Learn-Sustain") aimed at preparing undergraduate students in all majors to use their disciplinary knowledge and skills to contribute to the major societal challenge of creating sustainable communities. The initiative calls for faculty members from all six Georgia Tech colleges to develop courses and co-curricular opportunities that will help students learn about sustainability and community engagement and hone their skills by engaging in real-world projects with nonprofit, community, government, and business partners. Affiliated courses address various aspects of the Center's sustainable communities framework, which presents sustainability as an integrated system connecting

Integrating Sustainable Development into the Curriculum
Innovations in Higher Education Teaching and Learning, Volume 18, 125–143

ISSN: 2055-3641/doi:10.1108/S2055-364120200000018026

environment, economy, and society. This chapter reports on one engineering instructor's ongoing efforts that bring sustainability into the engineering classroom through sociotechnical project-based learning. This cornerstone design course is one of more than 100 Center-affiliated courses currently offered; the full set of Center-affiliated courses enrolls over 5,000 students per year across all six colleges. The sustainability activities introduced in the freshman design course pertain particularly to the Center's vision that all graduates of the institute, a majority of whom will graduate with engineering degrees, are able to contribute to the creation of sustainable communities and to understand the impact of their professional practice on the communities in which they work. A situated knowledge and learning pedagogical theory is used in the Center-affiliated course, where concept, activity, and context are involved in student learning to produce useable robust knowledge. The sociotechnical project-based teaching model with contextualized design problems is used to engage students throughout the course by utilizing computer-aided-design problems that incorporate sustainability within both individual and team projects. In this chapter, the authors present the pedagogical approaches to learning, strategies, and challenges for implementation and assessment of intervention activities, and data analyses of both student reflection data and pre- and post-survey data.

Keywords: Serve-learn-sustain center; sociotechnical project-based learning; situated cognition; external representation design-for-sustainability; freshman engineering; Real-world projects

INTRODUCTION

Integrating Sustainability in Education

A large and growing number of colleges and universities across the country have sustainability centers that engage in project work with community partners. The Sustainability Tracking, Assessment and Rating System program administered by the Association for the Advancement of Sustainability in Higher Education has more than 340 institutions, and each reporting institution must describe both academic and public-engagement-related sustainability programs. Sustainability in higher education goes far beyond campus recycling programs, and the intellectual influence of sustainability-related majors and programs is evident in the diversity of universities (public, private, large, and small), which boast leading sustainability programs, as the membership of groups such as the Sustainability Curriculum Consortium indicates. However, it is only within the last two decades that a critical mass of scholars has begun to connect sustainability-related initiatives and programs in higher education to both student learning and high-impact engagement or "systemic engagement" by universities. According to McNall, Barnes-Najor, Brown, Doberneck, and Fitzgerald (2015), systemic engagement facilitates "co-creating new understandings and co-designing new solutions to complex problems" (p. 4). Although the role of the higher education institutions

(HEIs) in strong, multi-stakeholder sustainability networks is critical to systemic engagement, undergraduate student learning within those HEIs is also crucial to configuring that new role for HEIs and the transformational engagement and teaching that accompany it.

Real-world project work on sustainability challenges can stimulate student growth from individuals set on solving an isolated problem to collaborators who have the competencies to work on complex projects with diverse stakeholders. Real-world project work on sustainability challenges facilitates bridging "sustainability" competencies, such as systems thinking, to those conventionally associated with professional or career-related training: team-work and communication, for example. Brundiers and Wiek (2017) point to six skills – teamwork, communication, preventative self-care, project management, "advanced continuous learning," and "effective stakeholder engagement" – as mostly untaught in many undergraduate degree programs and as critical to both specific sustainability project work and to professional success for students seeking sustainability careers (Brundiers & Wiek, 2017, p. 3). Educating students toward both nurturing the competencies and professional skills just described and advancing regional sustainable development goals requires forming partnerships that can impact classroom learning, strengthen collaborations across university silos, and support communities of practice beyond the university. It is in this context that the Center for Serve-Learn-Sustain (SLS) at Georgia Tech undertakes its work to equip students with the competencies to create sustainable communities and to advance sustainable development in the region.

Center for SLS at Georgia Tech

In January 2016, Georgia Tech launched a campus-wide academic initiative – the Center for SLS – aimed at preparing undergraduate students in all majors to use their disciplinary knowledge and skills to contribute to the major societal challenge of creating sustainable communities. The Center for SLS is our university's new quality enhancement plan (QEP), a key component of the University's reaffirmation of accreditation with the Southern Association of Colleges and Schools. The initiative collaborates with faculty in all six university colleges to develop courses and co-curricular opportunities that will help students learn about sustainability and community engagement and hone their skills by engaging in real-world projects with nonprofit, community, government, and business partners.

In addition to addressing course-specific learning outcomes, Center-affiliated courses are expected to address student learning outcomes (SLOs) related to sustainability and community engagement. Some of these SLOs are discussed later in the assessment section. Center-affiliated courses address various aspects of the SLS's sustainable communities framework, which presents sustainability as an integrated system connecting environment, economy, and society, and focuses especially on society – particularly social equity and community voice – to address this gap in technological education. Drawing on pedagogical research related to sustainable development and social justice, especially regarding engineering education, the framework asks faculty to help students understand how technology can assist or empower communities in their efforts to create places in which people and nature flourish, now and in the future (see Fig. 1).

Striving for systemic engagement, SLS's central pedagogical approach is project-based learning, and specifically, working on several priority issues with regional networks of partners to address sustainability challenges from community-based perspectives. As such, a key concern is helping engineering students focus on problem-framing before they jump into problem-solving. Specifically, we help students contextualize their projects so they understand technological solutions as one piece of a larger puzzle. As scholar Juan Lucena has suggested, there has been in the last 10 years a wellspring of scholarly interest and activity in problem- and project-based learning in what he broadly calls a "humanitarian" context. Lucena (2015) offers a critical vantage point on the legacy of humanitarianism in engineering by analyzing the traditional positing of engineers as emissaries of and authorities on "technology to 'relieve the suffering of these people'" (p. 16), pointing toward an imagined community of "people" who await salvation by American engineers. A "critical engineering education," as outlined by Lucena, is the one that equips students to be critical of such narratives and attentive to structural inequities, to seek projects that emerge from the community, to engage the expertise of community members, and to co-create, with local partners, structures that support an ongoing relationship, such that both the engineering "solution" and the collaboration between engineering students and community members can be sustained and nurtured.

SLS Center's approach to sociotechnical projects is distinguished by the wedding of an orientation toward justice and equity with an emphasis on technology as only one element of the dynamic interplay among society, technology, and nature. Service-learning scholar Tania Mitchell (2008), who reshaped the academic dialogue on service-learning and community-based learning, challenges the ways that "traditional service-learning" leaves structural inequality unexamined,

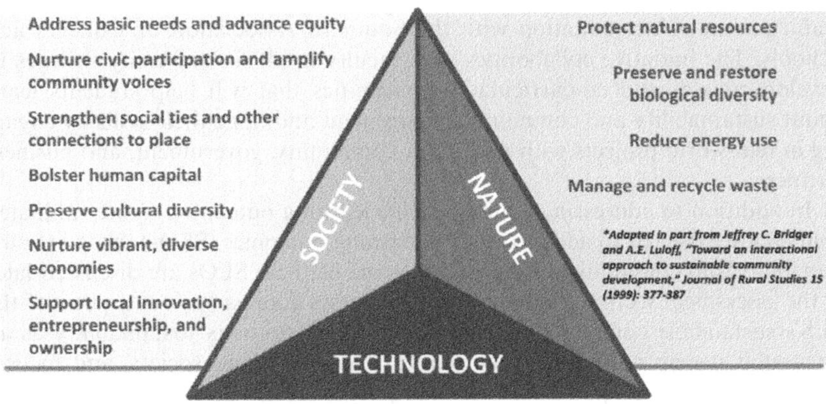

Fig. 1. How Can Technology Assist Communities in Supporting Society and Nature?

failing to guide students toward an understanding of its roots. What Mitchell (2008) calls "a social change orientation" and an aim to "redistribute power" distinguishes critical service-learning from traditional service-learning (p. 51 and 52). Drawing on this critical service-learning framework, the SLS Center aims to facilitate transformational learning for students as well as the foundations for longer-term relationships within communities.

To prepare engineering students with critical perspectives and deep context for local community-engaged work without sacrificing disciplinary rigor is a pressing challenge, and scholars of service-learning in engineering have taken it up fruitfully. Lima, Oakes, and Gruender (2006) emphasize the importance of robust institutional and intellectual architecture for service-learning projects in engineering. Envisioning the project as having outcomes at three levels – discipline/skill specific, "personal values," and "social systems and issues" – Lima et al. (2006) reinforce the importance of creating a trajectory for students that fosters their team-based, analytical, communication, and creative thinking skills in the context of a sustained and reciprocal relationship with a community partner (p. 309). When students are offered ways to engage in critical service-learning and are taught through critical engineering frameworks, they and their instructors and institutions are positioned to "broaden the narrow technical focus" of engineering education and see themselves as co-learners and co-teachers in project-focused relationships (Lucena, 2015, p. 16). Furthermore, when this form of collaboration is privileged, then relationships no longer have as their focus a given project or deliverable, but rather that those relationships endure and are nourished by their place in a broader network to effect deeper and more systemic local, national, or international change. Over the past three years, the SLS Center has worked closely with a diverse group of faculty to develop an experientially based pedagogical framework for engaging science and technology majors in sustainable communities' education – and to try it out in different classes.

This chapter presents that framework and reports on one engineering instructor's efforts to put it into practice by revising a core course in mechanical engineering. This course was one of more than 100 Center-affiliated courses offered during AY2016-17; the full set of Center-affiliated courses enrolls over 5,000 students per year across all six colleges – including engineering, computing, design, liberal arts, business, and sciences. The following section describes sociotechnical project-based learning interventions in integrating sustainability into a freshman-engineering design course.

INTEGRATING SUSTAINABILITY INTO A FRESHMAN-ENGINEERING DESIGN COURSE

Contextual Teaching and Learning and Situated Cognition

Situated cognition (Brown, Collins, & Duguid, 1989; Choi & Hannafin, 1995) emphasizes the importance of context in establishing meaningful linkages with learner experience and in promoting connections among knowledge, skill, and experience. Situated learning emphasizes higher order thinking rather than the acquisition of facts, encourages reflection on learning, and focuses on application rather than retention. Contextual learning (Khotimah & Masduki, 2016) engages students in meaningful, interactive, and collaborative activities that support them in becoming self-regulated learners. University training in problem-solving is primarily done using decontextualized textbook problems with a one-size-fits-all approach in exploring subject domain concepts. Teaching abstracted concepts as well-defined independent entities leads to learning with fixed meaning and immutable concepts. A situated approach to teaching and learning is implemented in the center-affiliated course, where concept, authentic activity, and context come together to guide students to produce useable, robust knowledge. Situated cognition leads to negotiable meaning and socially constructed understanding of the subject domain concepts.

Sustainability in Engineering Design Curriculum

Engineering design practice and education needs to change to address the sustainability challenges. Many pedagogical interventions have tried to address the problem of building sustainability values in the engineering curriculum. Date and Chandrasekharan (2017) provided a review of various classroom interventions related to sustainability in design course curricula: a team-based design project (Chau, 2007); problem-based learning approach (Dym, Agogino, Eris, Frey, & Leifer, 2005); and instructional laboratories in providing exposure to real-world problems (Feisel & Rosa, 2005). According to Date and Chandrasekharan (2017), engineering design education research indicates a two-fold gap that needs to be addressed to meet the challenge of integrating sustainability into the curriculum: (i) engineering graduates are technology-centered and are highly disengaged from the sociopolitical aspects of engineering and (ii) engineering graduates are not trained to situate themselves in problem contexts, particularly "wicked" real-world contexts, to independently frame problem requirements. In this work, a "sociotechnical project-based learning with context" teaching–learning model is used in a cornerstone design course to address these gaps.

About the Center-affiliated Course

Introduction to Engineering Graphics and Visualization is a freshman-engineering course in many universities. When Georgia Tech converted from quarter to semester curricula in 1999, the college of engineering created a three credit hour introductory engineering graphics course for undergraduates (Elrod & Stewart, 2002). In 2002, the course was revised using the "backward design" approach (Wiggins & McTighe, 1998) with formative and summative assessments in lecture and lab activities. In later years, project-based and learning-centered instructional approaches with creative ideation and sketching (Pucha & Utschig, 2012; Pucha et al., 2016) were introduced. Such approaches follow Kolb's learning model (Combs, 1976), which includes abstract conceptualization, active experimentation/application, concrete experience, and reflective observation. This is a core course for mechanical, civil, and aerospace engineering students offered in each

Fall and Spring semester with 2–4 instructors teaching 9–10 sections. One of the sections of this course is currently affiliated with SLS Center to integrate sustainability into the course curriculum.

Overview of Intervention Strategies: Sociotechnical Project-based Learning with Context

All students in this course learn: (i) creative design ideation; (ii) formal sketching techniques and orthographic projections; (iii) computer-aided-design (CAD) tools; and (iv) basics of design-for-manufacturing and 3D printing. Example prompts for student projects in regular sections of the course are shown in Appendices 1A and 2A for individual and team projects, respectively. Students typically chose coffee mugs for their individual projects and airplanes and other engineering structures for their team projects in the absence of any sustainability intervention. In the center-affiliated, experimental section of the course, sustainability themes are introduced in the individual and team projects, requiring students to explicitly address sustainability issues with their designs. The "sociotechnical project-based learning with context" teaching–learning model is implemented in the center-affiliated section through the following interventions:

(1) Individual projects that address human wasteful behavior of resources and environmental sustainability in product designs.
(2) Team projects with contextualized design problems that address social, environmental, and economic sustainability aspects in designing large engineering structures

Example prompts for student projects in the center-affiliated section are shown in Appendices 1B and 2B for individual and team projects, respectively. The sustainability-themed projects in the center-affiliated section are briefly discussed here.

External Representation of Design-for-Sustainability

Environmental Decision-making and Reasoning in the Context of Sustainable Development

Many of the decisions we make have implications for our environment, particularly those concerning natural resources and waste. When making everyday decisions, taking into account environmental factors can be both complex and challenging (Kiker, Bridges, Varghese, Seager, & Linkov, 2005). The human brain evolved to have two different modes of thinking, namely fast and slow (Kahneman & Egan 2011). Kahneman describes fast thinking (or System 1 thinking) as intuitive, automatic, unconscious, and effortless (Shleifer, 2012). Humans employ System 1 thinking in many life functions – we develop routines and heuristics to reduce the effort necessary to make everyday decisions. When we cannot remember whether we locked our door or much of the drive to work, it is evidence that our minds were in System 1 mode during those actions. Slow thinking (or System 2 thinking), on the other hand, is conscious, slow, controlled, and effortful (Shleifer).

Unfortunately, many environmentally friendly choices require us to use the slow mode of thinking, because the default option (the one that is easy to choose and carry out) is often wasteful in our world. For example, the default option at coffee shops is often a disposable cup and lid. Alternatives usually exist, such as bringing your own mug or consuming your beverage on site in a washable mug, but both of those options require a higher cognitive load. Bagozzi, Dholakia, and Basuroy (2003) identify four steps necessary to make an effortful sustainable decision if the sustainable choice is not the default: (1) a person first has to remember the decision (i.e., that reusable mugs are better for the environment than disposable ones); (2) they have to recognize the opportunity for decision enactment (i.e., when entering a coffee shop, they must realize that this is the time to ask for a reusable mug); (3) they have to activate and coordinate goal-directed behaviors (i.e., ask for a reusable mug or bring their own travel mug); and (4) they have to maintain the motivation through planning, monitoring the progress, overcoming impediments, and resisting temptations (i.e., put the travel mug in their bag every morning, do not buy coffee if they forgot, etc.). If any one of these steps is not observed, the person will fail to make the sustainable choice. When we break down the decision process like this, the complexity of the cognitive process becomes plain: the fact that a simple action like bringing a travel mug to work every day requires such attention and effort every time. This provides an explanation for why incentives alone (e.g., getting a discount on the coffee price for bringing a reusable mug) often do not lead to significant and consistent behavior change.

Sustainable External Representation Designs

The task environment of sustainable resource-use is highly unstructured and involves many uncoordinated and asynchronous actions as discussed above. To facilitate the process of sustainable decision-making, we can employ smart choice architecture (Thaler & Sunstein, 2009) to render making sustainable choices easier for people. Choice architecture is the deliberate design of a choice environment, where consumers are presented with the choices in a manner that makes the desirable choice easier. For example, one way to "nudge" people to eat healthier at buffet lines is to place the salad at the front of the line, and the unhealthy items at the end: the procession of the line will encourage eaters to fill up their plates with the healthy choices first, while they will have little room for the unhealthy options by the time they reach the end of the line. Choice architecture can be similarly employed and thus encourage people to make sustainable decisions. In product design, one of the ways to accomplish this is the smart use of external representation. Designing products with external representation (Chandrasekharan & Tovey, 2012) to improve the decision process using distributed cognition theories in structured problem-domains has been shown to reduce the cognitive load of individuals or groups. Application of distributed cognition in the design of consumer products with external representation to promote sustainability is still in its infancy as the design problem-domain is unstructured. When we design products with external representations specifically for sustainable resource-use, there are three objectives to accomplish: (1) make hidden information explicit in the design; (2) motivate people to make decisions that sustain resources; and (3) help people persist with the desired behavior by lowering the cognitive load involved in sustainability decisions.

In individual projects in the Center-affiliated course, students work on product designs with external representations, which promote sustainable resource-use and motivate people to make decisions and commit to behavior that sustains resources. Students use creative ideation and sustainability design concepts introduced in the lectures and prepare a preliminary set of visualization sketches describing the sustainability design with external representations. After feedback and approval from the instructor, students develop CAD models and working drawings, and then 3D print the parts. Fig. 2 shows examples of student work on external representation designs that promotes sustainable usage of appliances by making the implicit information explicit to the user with reduced cognitive load.

Humanitarian Design Projects with a Social Sustainability Theme

While university training in problem-solving is primarily done using decontextualized textbook problems, engineers are required to develop problem-defining skills in addition to problem-solving skills. In team projects that are sociotechnical,

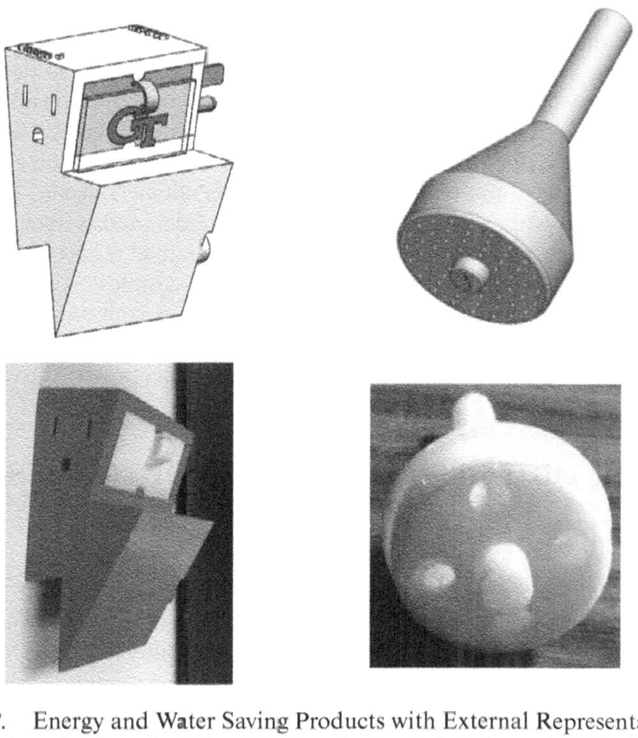

Fig. 2. Energy and Water Saving Products with External Representation Designs-for-Sustainability: Making the Implicit Information Explicit (CAD Models and 3D Printed Assembly).

Fig. 3. Rendered CAD Models of Student Work Samples on Sociotechnical Projects. (a) Multi-home and Low-cost Dehumidifiers. (b) Humanitarian Housing.

students are provided with more description of context, information on stakeholders involved and a list of available local resources. Up on conducting additional research on the chosen sociotechnical project from the database, students are required to design a simple, low-tech, low-cost, locally sourced product. For example, students (in teams of five) who have chosen the humanitarian housing project from the database give a detailed proposal to the instructor on alternative sustainable housing designs for refugees. In the proposal, students have to clearly identify various social sustainability aspects that are considered in the design including structural conditions meeting community needs (who benefits and who suffers), reducing imposed risks and harms posed by the environment and natural disasters, and enhancing human capabilities (Lucena, 2015). After approval, students work on design sketches, CAD models and system-level assembly of CAD models and functional designs with a detailed report for assessment. Fig. 3 shows rendered CAD models of student work samples on example project 1: Multi-home and low-cost dehumidifiers, which is a community project and example project 2: Humanitarian housing, which is sociotechnical case study design selected by students from the database of projects provided to them.

COURSE OBJECTIVES, LEARNING OUTCOMES, AND ASSESSMENT

Course Objectives and SLS Learning Outcomes

The Center-affiliated course has been constructed to foster a crucial mind-set in sustainability education: systems thinking. The interconnected and wicked nature of today's global environmental problems requires an ability to see the linkages and connections among elements of a system as well as among multiple systems, while maintaining an understanding that studying complex systems must happen holistically, with consideration of the context, the perspectives and the relationships of the actors, and the system's boundaries. Systems thinking provides an understanding of a system by examining the linkages and interactions between the elements that comprise the whole system. Sustainability systems-thinking skills include: (1) identifying dynamic relationships among ecological, social, and economic factors of sustainability; (2) understanding the influence of context; and (3) evaluating how design decisions influence the sustainability of communities.

To develop this mind-set, the course pursues higher order learning outcomes on Bloom's revised taxonomy (Krathwohl, 2002). Fig. 4 demonstrates how the lower three levels of the taxonomy (remember, understand, and apply) ask students to consider the elements of systems individually, whereas to achieve the upper three (analyze, evaluate, and create) students must "put things together." At the "analyze" level, students are exploring the interconnections within a system and studying how alterations to one element change the behavior of the whole. At "evaluate," students are able to compare and contrast alternatives, such as two proposed solutions to a design problem. By the end of the course, we ask students to "create": they are designing a working system of their own, a new product that takes into account the economic, societal, and environmental systems, and formulates a new solution using their previous knowledge. Some of the course-learning objectives align with the learning outcomes of the SLS Center. These learning outcomes are presented in Table 1 along with the assessment instruments.

Assessment

Georgia Tech is well known for its disciplinary excellence. However, as shown in Table 2 there was a disparity in student assessment of preparedness with respect to various skills from Georgia Tech's 2012 Baccalaureate Alumni Survey. More than 75 percent of Georgia Tech graduates rated their disciplinary skill preparation as high. Notably, fewer gave their "interskill" preparation a high rating with regard to

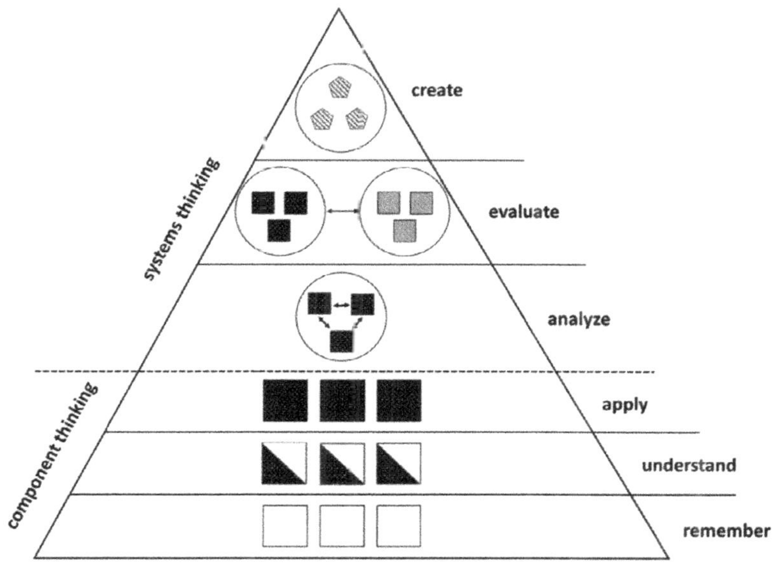

Fig. 4. Systems Thinking, Bloom's Taxonomy.

Table 1. Student Learning Outcomes (SLOs) and Assessment Instruments.

SLOs	Assessment Instruments
Students will be able to identify relationships among ecological, social, and economic systems	Pre and post surveys (word cloud analysis)
Students will be able to evaluate how decisions impact the sustainability of communities	Open-coding and comparison of Post-activity reflection on sustainability intervention in design projects (thematic analysis)
Students will be able to describe how they can use their discipline to make communities more sustainable	GT-SLS Rubrics for grading pre and post surveys, post-activity reflections and student design projects using SLOs

Table 2. Student Assessment of Preparedness from Georgia Tech's 2012 Baccalaureate Alumni Survey.

Statement: To What Extent Did Georgia Tech Prepare You to	% Saying Well or Very Well Prepared
Disciplinary skills	
Identify, formulate, and solve problems in your discipline	87.6
Use techniques, skills, and tools needed for the practice of your discipline	76.8
Seek out new information needed for the practice of your discipline	75.5
Interskills	
Function on multi-disciplinary or cross-functional teams	68.6
Effectively resolve interpersonal conflict within a group or team	51.7
Understand the social and cultural impact of your professional practice	**44.3**
Understand the environmental impact of your professional practice	**40.4**

Source: 2012 Baccalaureate Alumni Survey.

effective work in a team or understanding the environmental, social, and cultural impact of their professional practice. This disparity provides one of the motivations for the Center for SLS as part of a new QEP. The "sociotechnical project-based learning with context" model used in the center-affiliated section of the course aims to assess: (1) students' perceptions of the intervention; (2) students' learning of sustainability system-thinking skills during the ideation, implementation, and evaluation stages of the engineering design process; and (3) students' understanding of the social, cultural, and environmental impact of professional practice.

Methods and Results

Student Survey

Students in the center-affiliated, sustainability-focused section were given a survey intended to capture their standing on the learning outcomes of interest, as well as their general perceptions of the sustainability intervention itself. The survey was implemented using a pre–post design, with the student perception items being given only in the post, as there is no student experience with the intervention at the pre-survey administration. Comparison of pre- and post-student responses on the items related to the learning outcomes allows for an examination of the extent to which students' understanding of the relationships among various systems and the drivers of sustainability changed over the course of the intervention.

The survey was administered online using the Survey Monkey platform, with the pre-survey being administered roughly two weeks into the course, and the post-survey being given during the last week of the course before final exams. The survey took roughly 20 minutes for students to complete and was administered online with survey links e-mailed to students. Across survey versions, there were 6–11 items on the pre-survey and 9–11 items on the post-survey, including Y/N, multiple choice, and open-ended items. Survey items were designed to elicit student responses reflective of both their standing on each of the three learning outcomes presented in Table 1 and their perceptions of the sustainability intervention. This survey is part of the larger sustainability initiative, taking place at the university, and was used in many other sustainability-focused courses.

Fig. 5 shows an example of pre- and post-survey analysis results on student responses to the survey question "rate your understanding of the relationship between ecological, social and economic systems." The affiliated section had 40 students: 19 students responded to the pre-survey and 16 students responded to the post-survey.

Fig. 6 shows an exemplary word cloud representation of students' pre- and post-survey responses to the survey question: "In your own words define sustainability." As can be seen from the word cloud representation at the pre-survey time point, the majority of students define sustainability as relating primarily to environment. In the post-survey word cloud representation, one can notice that more students seem to recognize that product designs with a sustainability focus are

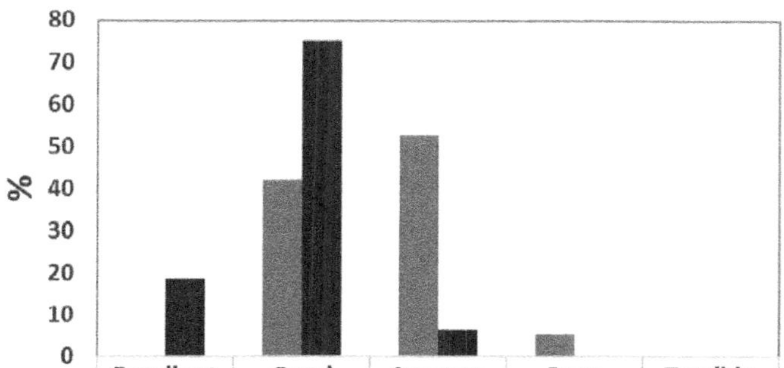

	Excellent	Good	Average	Poor	Terrible
Pre	0	42.11	52.63	5.26	0
Post	18.75	75	6.25	0	0

Fig. 5. Pre- and Post-survey Response Analysis.

Fig. 6. Pre- and Post-survey Response Analysis: Word Cloud Representation. (a) Pre-survey Response. (b) Post-survey Response.

associated with ecological, social, and economic factors. Sample pre- and post-survey responses are shown below.

Sample pre-survey response: "Designing products so that they last with time while doing minimal harm to the environment."

Sample post-survey response: "Sustainability is the incorporation of ecological, social, and economic factors meant to ensure that a product or process benefits all at the expense of none."

The limitations of self-report surveys of this nature in reflecting student learning should be noted, in that students may be limited in their ability to assess their true understanding of complex and ill-defined subject matter. This limitation is somewhat addressed in this work through including the analysis of student work products using thematic analysis conducted on the open-ended student responses for each of the three sustainability conceptual understanding items. The resultant themes and illustrative quotes from student responses are presented below.

Student Reflections

Engaging students in reflection activities can be an important learning tool. It gives students an opportunity to reflect on their experiences and think critically about what they have learned. As both a self-reflection activity and an additional source of data on their understanding of sustainability within the context of their specific course and project, students were asked to respond to a series of reflection questions as part of their individual design projects on external representation design-for-sustainability (described in the "Center for SLS at Georgia Tech" section). The reflection questions, which were a required component of the project report, are as follows:

(1) Discuss at least three ways by which your product design promotes sustainable usage of resources.
(2) What is your understanding of the relationship among environmental, social, and economic systems?

(3) Think about your design project and discuss how the 3Ps (1) social sustainability (people); (2) environmental sustainability (planet); and (3) economic sustainability (profit) are reflected in your project design?
(4) What did you learn by creating this product? Describe how your engineering discipline can make communities more sustainable?

Student responses to these prompts from the Fall 2018 semester were analyzed using thematic analysis, with a focus on evidence of students' level of understanding of the interconnectedness of social, economic, and environmental systems, as well as the drivers of sustainable communities and how students' disciplines can support sustainability in communities. The results of the thematic analysis of student reflections, including the eight major themes identified and the frequency of each in the set of student responses are provided in Table 3. An illustrative quote from a sample student reflection reads as follows: "I have learned that a 'green' product is not a sustainable product because sustainable means that it has economic and social implications instead of just being created from recycled material."

CONCLUSION

Several programs currently support Georgia Tech SLS Center's vision of students drawing on their disciplinary expertise to help "create sustainable communities." All of these programs allow students to explore different avenues of engagement and provide opportunities for them to connect course-based projects to broader, multi-stakeholder initiatives. Affiliated courses address various aspects of the center's sustainable communities framework, which presents sustainability as an integrated system: connecting environment, economy, and society (see Fig. 1).

Table 3. Individual Project Post-activity Reflection Results.

Student Reflection Results	Frequency
Three ways product design promotes sustainability and relationship among 3Ps (planet, people, and profit)	
Value-attitude	21
Help-allowed-understand	16
Social-environment-community-context-human	22
What did you learn by creating this product?	
Design-innovate-solve	45
Skill-experience-creative-learned	33
Active-apply-think-improve-focus	24
Ideation methods	9
How engineering discipline can make communities more sustainable?	
Challenged-appreciated-impact	26

99 responses, 602 parsed items coded.
97.5% of comments were positive or neutral.
Additionally,
1[illegible] student comments on *lack of clarity* about project expectations or instructions.
3 comments about this type of creative work being *hard.*
7 comments that they *enjoyed practicing* the ideation methods.

In this chapter, one engineering instructor's ongoing efforts to bring SLS aspects into the engineering classroom through sociotechnical project-based learning is presented. Assessment with pre- and post-course surveys of students' perceptions on interventions demonstrate the changing understanding of sustainability that students develop in the course. Through written reflections, students demonstrate their grasp of basic elements of SLS thinking. Further analysis of student responses to the interventions, as well as analysis of their work products using the rubrics developed as part of the university's larger sustainability initiative, is needed to quantify students' relative standing with respect to various learning outcomes.

REFERENCES

Bagozzi, R. P., Dholakia, U. M., & Basuroy, S. (2003). How effortful decisions get enacted: The motivating role of decision processes, desires, and anticipate emotions. *Journal of Behavioral Decision Making*, *16*, 273–295.

Brown, S. J., Collins, A., & Duguid, P. (1989). Situated cognition and the culture of learning. *Educational Researcher*, *18*(1), 32–42.

Brundiers, K., & Wiek, A. (2017). Beyond interpersonal competence: Teaching and learning professional skills in sustainability. *Education Sciences*, *7*(1), 39.

Chandrasekharan, S., & Tovey, M. (2012). Sum, quorum, tether: Design principles underlying external representations that promote sustainability. *Pragmatics & Cognition*, *20*(3), 447–482.

Chau, K. W. (2007). Incorporation of sustainability concepts into a civil engineering curriculum. *Journal of Professional Issues in Engineering Education and Practice*, *133*(3), 88–91.

Choi, J., & Hannafin, M. (1995). Situated cognition and learning environments: Roles, structures, and implications for design. *Journal of Educational Technology Research and Development*, *43*(2), 53–69.

Combs, A. W. (1976). Fostering maximum development of the individual. In W. Van Til & K. J. Rehage (Eds.), *Issues in Secondary Education*, (pp. 65–87). Chicago, IL: University of Chicago Press.

Date, G., & Chandrasekharan, S. (2017). Beyond efficiency: Engineering for sustainability requires solving for pattern. *Engineering Studies*, *10*(1), 1–26.

Dym, C. L., Agogino, A. M., Eris, O., Frey, D. D., & Leifer, L. J. (2005). Engineering design thinking, teaching, and learning. *Journal of Engineering Education*, *94*(1), 103–120.

Elrod, D., & Stewart, M. D. (2002). Assessing student work in engineering graphics visualization course. In *Proceedings of the 2002 American Society for Engineering Education annual conference & exposition, Montreal, Canada*.

Feisel, L. D., & Rosa, A. J. (2005). The role of the laboratory in undergraduate engineering education. *Journal of Engineering Education*, *94*(1), 121–130.

Kahneman, D., & Egan, P. (2011). *Thinking, fast and slow* (*Vol. 1*). New York, NY: Farrar, Straus and Giroux.

Khotimah, R. P., & Masduki, M. (2016). Improving teaching quality and problem solving ability through contextual teaching and learning in differential equations: A lesson study approach. *Journal of Research and Advances in Mathematics Education*, *1*(1), 1–13.

Kiker, G. A., Bridges, T. S., Varghese, A., Seager, T. P., & Linkov, I. (2005). Application of multi-criteria decision analysis in environmental decision making. *Integrated Environmental Assessment and Management*, *1*, 95–108.

Krathwohl, D. R. (2002). A revision of Bloom's taxonomy: An overview. *Theory into Practice*, *41*(4), 212–218.

Lima, M., Oakes, W. C., & Gruender, J. L. (2006). *Service-learning: Engineering in your community*. Wildwood, MO: Great Lakes Press.

Lucena, J. (2015). Bridging sustainable community development and social justice. In S. Christensen, C. Didier, A. Jamison, M. Meganck, C. Mitcham, & B. Newberry (Eds.), *International perspectives on engineering education*. Engineering in Practice and Context (Vol. *1*, pp. 225–247). Basel, Switzerland: Springer.

McNall, M., Barnes-Najor, J., Brown, R., Doberneck, D., & Fitzgerald, H. (2015). Systemic engagement: Universities as partners in systemic approaches to community change. *Journal of Higher Education Outreach and Engagement*, *19*(1), 7–32.

Mitchell, T. (2008). Traditional vs. critical service-learning: Engaging the literature to differentiate two models. *Michigan Journal of Community Service Learning*, *14*(2), 50–65.

Pucha, R. V., & Utschig, T. T. (2012). Learning-centered instruction of engineering graphics for freshman engineering students. *Journal of STEM Education: Innovations and Research*, *13*, 24.

Pucha, R. V., Utschig, T. T., Newton, S. H., Alemdar M., Moore R., & Noyes C. R. (2016). Critical and creative thinking activities for engaged learning in graphics and visualization course. In Proceedings of the 2016 *American Society for Engineering Education annual conference & exposition, New Orleans, LA, USA*

Shleifer, A. (2012). Psychologists at the gate: A review of Daniel Kahneman's thinking, fast and slow. *Journal of Economic Literature*, *50*(4), 1080–1091.

Thaler, R. H., & Sunstein, C. R. (2009). *Nudge: Improving decisions about health, wealth, and happiness*. New York, NY: Penguin.

Wiggins, G., & McTighe, J. (1998). Understanding by design. In *Assessment for learning conference*, Ljubljana. Alexandria, VA: ASCD.

APPENDIX 1. INDIVIDUAL PROJECTS

A. Prompt for Product Design in Regular Course Section

Using ideation, sketching and 3D CAD design the creative and unique souvenirs for home and office use or décor for Georgia Tech alumni.

B. Prompt for Product Designs with External Representations Promoting Sustainable Resource-use in Center-affiliated Section

Product design with external representations promoting sustainable resource-use can motivate people to make decisions that sustain resources and persist with this behavior. Designs with external representations promoting sustainability will lower the cognitive load involved in sustainability decisions. Design creative and unique appliances that promote sustainable use of natural resources for home and office use of Georgia Tech community.

APPENDIX 2. TEAM PROJECTS

A. Prompt for Design on Engineering Structures in Regular Course Section

Use creative ideation, sketching and 3D CAD tools to design, model and assembly of large engineering structures. Create animations reproducing functionality of designed engineering structure.

B. Prompts for Humanitarian Design Team Projects with Contextualized Problem Definition in Center-affiliated Section

Using the context, structural condition and other social and economic constraints provided in the database of projects: design, model, and assemble the chosen engineering structures. Examples of a community project and a humanitarian design project are briefly described below.

Example Project 1: Design low-cost dehumidifiers as a potential solution for reducing public health risks associated with mold and mildew in the Proctor Creek watershed, Atlanta. The water vapor removed from air in homes could potentially be reused for gardening and dehumidifiers that utilize solar energy may not have a huge impact on energy bills. Alternative low-cost multi-purpose de-humidification approaches and designs are also encouraged.

Example Project 2: United Nations High Commissioner for Refugees (UNHCR) estimates that 10 percent of the world's refugee population lives in tents, many of which do not provide adequate insulation from the heat or the cold and offer no electricity. Current refugee camps are more likely to hold canvas ridge variety tents, similar to a traditional scout tent, and lightweight hoop tents, none of which is particularly durable or mobile under unexpected situations. With the average stay in a camp being around 12 years, these current conditions must be challenged with new, innovative shelter solutions that can adapt to as many environmental, social, and economic conditions as possible while maintaining their purpose.

CHAPTER 9

INTEGRATING SUSTAINABLE DEVELOPMENT INTO THE CURRICULUM: ENACTING "SCALAR SHIFTING" IN ESD COMPETENCIES

Paul Benneworth, Renze Kolster, Martin Stienstra, Laura Franco Garcia and Ben Jongbloed

ABSTRACT

There is an increasing interest in integrating sustainable development into higher education curricula to increase young graduates' agency in addressing sustainable development goals (SDGs). Education for sustainable development (ESD) involves raising awareness of opportunities to create local solutions, a willingness or desire to construct those solutions, and the organizational skills to implement these solutions in context. As these courses are integrated into academic curricula, students must learn practical-ends driven skills in ways compatible with existing academic standards frameworks oriented toward theoretical understanding. These can lead to very different pedagogical orientations (theoretical and practical), one reason that could explain the relatively limited uptake of ESD within higher education to date. The authors develop a model by which a single educational experience could help to bridge between these two orientations. The authors use a single study of an example of student volunteer projects where students spend 2–4 months working on a knowledge

Integrating Sustainable Development into the Curriculum:
Innovations in Higher Education Teaching and Learning, Volume 18, 145–161

ISSN: 2055-3641/doi:10.1108/S2055-364120200000018027

transfer project to the global south, oriented toward solving the SDGs. The authors reflect on tensions, problems, and solutions in producing graduates oriented to tackling urgent contemporary societal issues, while gaining valuable personal development experience.

Keywords: Education for sustainable development; sustainable development goals; grand challenges; bachelor-level education; HE minor programs; academic volunteering programs, university third mission; university societal mission

INTRODUCTION

There is an increasing interest in systematically integrating sustainable development into higher education curricula to increase young graduates' sense of autonomy and agency in addressing sustainable development's complex challenges (Dale & Newman, 2005). This education for sustainable development (ESD) is not merely about providing knowledge of the challenges in delivering sustainable societies but in raising awareness of opportunities to create local solutions, a willingness or desire to construct those solutions, and organizational skills to implement these solutions in context (Molderez & Fonseca, 2018). The 2009 Bonn Declaration noted that:

> ESD emphasises creative and critical approaches, long-term thinking, innovation and empowerment for dealing with uncertainty, and for solving complex problems. ESD highlights the interdependence of environment, economy, society, and cultural diversity from local to global levels, and takes account of past, present and future. (United Nations Educational, Scientific and Cultural Organization (UNESCO), 2009, p. 2)

As ESD is integrated into academic curricula, students must learn practical-ends driven skills in ways compatible with academic standards frameworks oriented toward accrediting theoretical understanding (Quality Assurance Agency (QAA), 2014). This chapter foregrounds these tensions in integrating ESD into curricula, asking the research question of whether learning communities can be constructed spanning "classroom to local project," where practical experiences are coupled to theoretical learning outcomes. We contend that although sounding relatively straightforward, the divergence in pedagogic orientation between abstract-theoretical and specific-practical is one reason that could explain ESD's relatively limited uptake within higher education to date.

We develop a model by which a single educational experience could help to bridge between these two orientations. We also highlight the key issue here being transitions between the problem's different scales and relate that to the development of key ESD competencies. We explore this conceptual framework with a single study of a student volunteering minor where students spend 2–4 months working on a knowledge transfer project to the global south, simultaneously researching how that project contributes to the sustainable development goals (SDGs). We use qualitative information based on submitted course material (including the research report, a reflective report, and testimonials) and feedback from course participants. This allows our chapter to reflect on tensions, problems, and solutions in ensuring universities produce graduates well-oriented to tackling urgent contemporary societal issues, while also gaining valuable personal development experience and career perspectives.

THE TENSIONS OF BLENDING PRACTICAL AND ACADEMIC EDUCATION

SDGs and Learning About Delivering Change

The current understanding of ESD is shaped by the 2009 Bonn Declaration, itself an offshoot of the decade of "Education for Sustainable Development" seeking to use education to drive the millennium development goals (United Nations Educational, Scientific and Cultural Organization (UNESCO), 2012). The ambitions for ESD are extensive, to translate the education sector into a mechanism for delivering equality and social justice by empowering people for change. The ambition is that ESD equips learners to integrate different kinds of knowledge into different constellations in order to solve place-specific local challenges. United Nations Educational, Scientific and Cultural Organization (UNESCO) foreground the importance of a multi-perspectival approach to educating learners to create these new knowledges, highlighting eight different perspectives (i.e., scientific, historical, geographical, human rights, gender equality, values, cultural diversity, and sustainability).

A central component of ESD is empowering and inspiring people to action, creating actionable knowledge and then using that knowledge to improve local circumstances (Leicht, Heiss, & Byun, 2018). As Leicht et al. (2018) observe, ESD has a dual nature, not only improving understanding but also driving action and change. This has been addressed through the notion of creating "competencies" for ESD, equipping learners to participate meaningfully in local processes and to take control of their own environments (Shephard, Rieckmann, & Barth, 2018). As Shephard et al. (2018) noted, Wiek, Withycombe, & Redman (2011) contend that there are five key ESD competencies, namely systems thinking, futures thinking, normative thinking, action-oriented thinking, and interpersonal competencies, in 2016 adding the competency of integrating those five competencies to solve the real problems of sustainable development. Indeed, the addition of this sixth competency highlights a substantive problem in ESD, namely the gap between diagnosing a problem, planning a suitable course of action, and delivering that action in practice (Wiek et al., 2016).

ESD's roots lie in the original Rio Agenda 21 summit (1992): the question remains as to why three decades later does ESD remain as a relatively marginal educational concept despite the obvious urgency of the challenges and problems which it addresses. From a higher education perspective, implementing a solution as part of the learning experience runs into questions of fairness of assessment, notably in assessing practical learning focused on delivering system change. A particular ESD project's success or otherwise depends far more on the wider project environment rather than on the learner's direct intervention. This raises a question of the status of a practical experience or placement within the overall learning experience, and how to give credit to learners who are independent on sheer luck nor the munificence of the political environment within which they are operating.

In this chapter, we draw an analogy between ESD and entrepreneurial education. A key problem in entrepreneurship education is a strong split between theory-driven education and action-driven education (Benneworth & Osborne, 2016). In theory-driven courses, entrepreneurs are an object of study, whereas in action-driven education, creating a new enterprise is emphasized. It has proven persistently difficult for entrepreneurship educators to find ways to blend the theory-centered and action-centered elements of education. Indeed, although theory-centered courses typically bear study points, many enterprise courses fall outside formal accreditation mechanisms, and risk becoming detached from their theoretical foundations, thereby losing their connections to higher education (Fayolle, Verzat, & Wapshott, 2016). This provides a plausible diagnosis of the contemporary ESD problem, how to create competencies for action not assessed on those actions' success, thus allowing ESD to remain conceptually coupled to universities' theoretically centered knowledge activities within universities.

Creating a learning linkage between scales: from practical project intervention to SDG solution.

A key challenge is the conceptual gulf between the SDGs and their manifestations in practice. The SDGs can be regarded as a "washing list" of everything afflicting humanity, grouped under 17 headings in an attempt to make those problems knowable at the international policy level. There is no one-to-one correspondence between the SDGs, the various problems that constitute a single SDG, and potential solutions to those problems, and their symptoms. Focusing on the SDGs as the object of study can encourage the oversimplification of complex problems and the proposition of generic, unspecified solutions ("we need a new governance mechanism to solve SDG10[1]"). Conversely, focusing on immediate problems can lead to studying symptoms rather than identifying necessary changes to transform behaviors ("we can solve the hunger challenge by giving people food").

Consider the example of Sustainable Development Goal 2, End Hunger (UNESCO, 2019). This consists of five subgoals (access to food, ending malnutrition for young and elderly people, raising agricultural productivity, improving resilience of agricultural landscapes, and ensuring agricultural biodiversity) along with three structural changes (increasing rural investment, eliminating unfair trade practices, and creating transparent agricultural commodity markets). These problems manifest themselves in complex, interrelated ways in particular places: lower agricultural landscape resilience and unfair agricultural commodity markets can create food insecurity undermining food access, whereas increasing rural investment is intimately bound up with another SDG, that of improving female emancipation (Juliani, Tamilis da Silva, Cunha, & Benneworth, 2017). Any abstract analysis of these situation risks remain divorced from the practical level, making recommendations not themselves directly actionable. Conversely, there is a risk that any practical analysis of a situation of food insecurity fails to understand the wider causal relationships that produce these symptoms.

One way to bridge between the conceptual and practical elements is to use theory to diagnose why a particular situation exists the way it does and to ask whether a particular local intervention addresses the underlying causes. From a pedagogic perspective, the act by the student of *making that diagnosis* (from abstract to particular) would deliver three of Wiek et al. (2011)'s ESD competences, which are the following:

- *Systems thinking*: identifying how global systems, for example, agricultural commodity markets impact on particular places.
- *Futures thinking*: making a specific diagnosis of how the future can be improved by addressing an underlying mechanism, for example, knowledge sharing between farmers.
- *Normative thinking*: identifying a particular social conditions (such as food insecurity) as being "unjust" and therefore require challenging.

What is absent here in terms of Wiek et al.'s competence set are: (a) action-oriented thinking, (b) interpersonal skills, and (c) the meta-competency of integrating these together. We here see the potential for a link to a practical project to:

- attempt to undertake such a project (demonstrating interpersonal skills) and
- use that diagnosis to inform behavior within the projects (action-based knowledge).

The final meta-competency is that of integrating these five competencies by taking the learning experiences derived from the practical project and reflecting on what this then says about the solution of the SDG more generally. We therefore propose the following model (see Fig. 1) for integrating the ESD learning

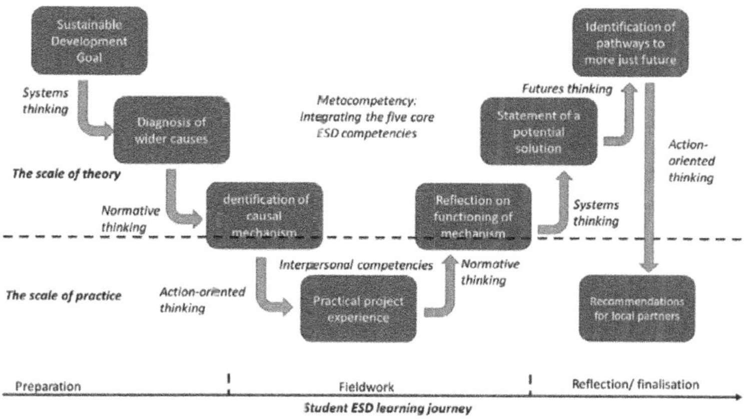

Fig. 1. Toward a Multi-scalar Model of the Student Learning Journey Associated with University ESD.

competencies in ways that bridge between theoretical and practical learning approaches.

METHODOLOGY

Operationalization

In order to ask whether this model does indeed lead to the development of the desired competencies, we undertake an exploratory study, reflecting the exploratory nature of the model presented in Fig. 1. The model suggests that the necessary condition for bridging between theory and action in ESD can be delivered if there is a *scalar shift* that has two characteristics, namely: (a) a shift from the abstract SDG to a practical problem informed by theory and (b) that practical problem then provides the basis for testing whether a particular mechanism can help address one of the causes underlying the prevalence of that SDG. We ask whether those competencies have been developed, and whether they correspond to scalar-shifting processes (from the general→specific→general) in the model depicted in Fig. 1. In "scalar shifting," we here refer to the fact that working at one scale creates knowledge within the student that affects material practices at the other (e.g., that diagnosing that food insecurity is not caused by a lack of food but a lack of capacity to secure access to resources, and then using that insight in the practical project).

We do this with reference to one intervention within a single university context, that of a minor program called "Crossing Borders" offered at the University of Twente (UT). That minor program offers all bachelor students the opportunity to work on a project to contribute to solving an SDG (see below). As part of that minor, all participants had to produce a study reflection report, in which they reflected upon the whole learning experience. The reflection reports also provide a means for the teachers to attune their teaching approaches for subsequent cohorts and to analyze the extent to which the students meet the formal learning goals. We have assembled 190 reflection reports from four cohorts of students following the minor, and these form the basis for the data presented in this chapter.

We then analyzed those reflection reports to identify where students had presented material for this scalar-shifting processes (from the practical to the abstract or vice versa) and identified a total of 56 extended comments that could be related to our conceptual model. We sought out comments where the students did not present bald statements but gave corroborating evidence that described their behavior during the volunteering placement. In all, 34 of these related to the experiences on the module, 13 were reflections on the learning process, 4 on the nature of the practical experience, and 1 generic comment. We have organized the relevant data under the six competencies, and then attempted to produce a stylized overview of how these processes functioned. All comments have been anonymized.

On that basis, we are then able to produce an iteration of our conceptual model that specifically addresses the issue of scalar shifting, and we are able to identify a set of pedagogic challenges for delivering ESD. We are then able to identify how ESD can better be implemented in higher education contexts in ways that help students create theoretically rooted actionable knowledge that contributes to delivering the SDGs. We acknowledge that the relatively "thin" nature of our evidence base makes this a slightly speculative piece of work and emphasize that this chapter seeks to provide insights for the further development of ESD rather than making a definitive evaluation of the ways in which the students did or did not develop competencies.

INTRODUCTION TO THE CROSSING BORDERS CASE STUDY

We here report findings from a single minor program at the UT, "Crossing Borders." UT is a small technical university in the East of the Netherlands with 11,000 students studying primarily science & engineering and social science disciplines. The university introduced the minor structure in 2000 as a response to the Bologna higher education reforms where the Netherlands switched to a system of bachelor and master degrees. In 2012, UT overhauled its curriculum around the so-called Twente Education Model (TEM) where each bachelor degree is formed from 12 modules, and for each quartile, students exclusively follow that module. All minor courses correspond to quartiles 9 and 10, the first semester of the student's third year, immediately before the dissertation modules, which lead to graduation at the end of the third academic year. Minors are not compulsory; students may also undertake an exchange program (e.g., ERASMUS) or follow an alternative placement. Students who undertake a placement abroad are eligible for a travel grant via the Twente Mobility Fund.

Crossing Borders was launched in 2015 as the first TEM cohort started their minor programs. The program's rationale is to provide an ESD experience where students work on a project (typically in the Global South) that contributes to addressing a named SDG. The module's nomenclature alludes to the various kinds of borders that students must cross to successfully pass the module, including geographical borders, the border between technological and social innovation, cultural borders, team working, and multidisciplinarity. Different variants allow students to choose a duration of one or two quartiles: a long placement (two quartiles), a short placement (one quartile), and a one quartile module by going on a study tour abroad, or doing a project without leaving the Netherlands (working on a "virtual" project) (see Fig. 2).

Students carry out an external placement registration in quartile 8 and start searching individually (but assisted) for a host organization and study topic. A host organization is a social enterprise taking volunteers who deliver services, typically a nongovernmental organization, a hospital or school, or a university. Each student is allocated an academic supervisor assisting with practical project elements, particularly preparing the research plan and two final reports.

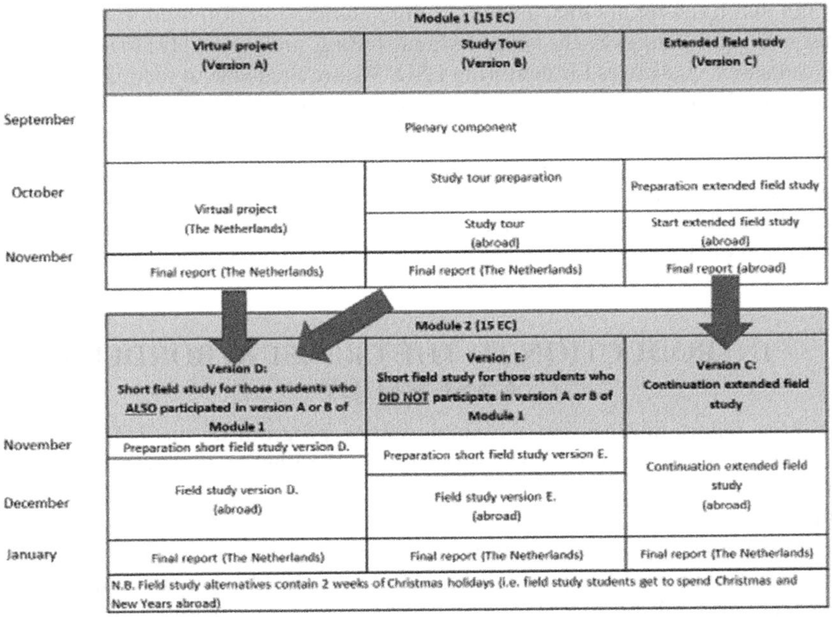

Fig. 2. A Schematic Overview of the Throughput in Crossing Borders.

Table 1. The Location of Foreign Placements in Crossing Borders.

Indonesia	18%	Australia	3%
South Africa	10%	Ghana	3%
Thailand	6%	Kenya	3%
Tanzania	5%	Nepal	3%
Costa Rica	4%	Uganda	3%
Peru	4%	United States	3%
Suriname	3%	Other[a]	34%

Source: Authors' own calculations; $n = 222$.

[a]Three students (2%) visited Cambodia, Vietnam, Argentina, Bonaire, Namibia, and Sri Lanka; two students (1%) visited Brazil, Canada, Madagascar, Malawi, Singapore, Bolivia, Curacao, India, Jamaica, Malaysia, Mexico, Palau, and Philippines; and one student (<1%) visited Bali, Belize, Chile, Gambia, Greece, Guadeloupe, Guinee Bissau, Mozambique, Netherlands, New Zealand, Panama, Poland, Portugal, Romania, South Korea, Spain, Turkey, and United Kingdom.

Table 1 provides an overview of destination countries (this is based on students who recorded a clear destination in the database).

Each variant begins with several weeks' classroom-based study in Twente focusing on research skills, understanding SDGs, working in cross-cultural contexts, social innovation, and other theoretical insights (the Grand Challenges, Human Capital, so-called "21st century skills," and Tragedy of the Commons). In this phase, students undertake two tasks in groups formed by the teachers to mix students from different courses, a short poster presentation and a short paper, each relating to an SDG not necessarily related to their overall project.

Students are allocated an SDG for these tasks so that not every group works on the same topic.

The students then undertake (individually or in pairs) a practical volunteering project within a host organization, and research exploring how a particular SDG is being addressed in practice. Typical tasks undertaken for the host organization for local community benefit include teaching English, contributing to marketing strategies, developing communications campaigns, and participating in social credit and community investment schemes. A student research project associated with the teaching English placement might be asking what kind of motivations adult learners of English in rural Peru have for part-time study. Each student prepares a country report, a research plan, a final research report, a reflection report, and a testimonial paper. In the four years, 249 students completed a variant involving working abroad, 144 were longer placements, 28 were short studies following the virtual project or a study tour, and 77 completed the short study in quartile 10 only. The Twente Mobility Fund provided €114,000 for student support, along with €10,000 from Holland Scholarships, and €2,000 from ERASMUS funding, a total "cost" to the university of €126,000, or roughly €30,000 per cohort.

To illustrate this (see Fig. 3), we provide a hypothetical example of a single assignment oriented toward SDG2, Zero Hunger, for students who agree in July to undertake a marketing placement for the Great Lakes Dairy Development Cooperative, that organizes dairy processing and refrigeration for dairy farmers in East Africa.[2] In their poster presentation and short paper, prepared in Twente in October, they diagnosis the problem is food insecurity with farmers locked into precarious situations by an inability to invest in their own businesses, potentially derived from gendered household income arrangements. The student pair travel in November to East Africa to develop a marketing plan for GLDD. The research project they undertake focuses on gender barriers to investment and explores

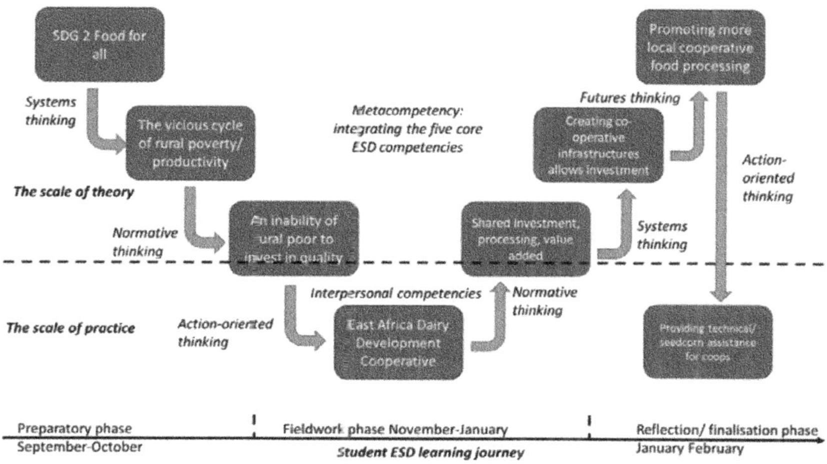

Fig. 3. An Archetypical Crossing Borders Project.

whether GLDD helps female farmers to develop independent income for their household context. They find that the cooperative model is effective in helping to provide more women farmers with more access to investment finance and to raise their business productivity. They recommend in their reports more gender-specific mentoring and gender activity within GLDD as a means to address this vicious circle of food insecurity and SDG2, No Hunger.

SCALAR-SHIFTING PROCESSES IN THE ESD'S SIX COMPETENCIES

We explore how developing the identified competencies was associated with scalar-shifting processes, focusing on six competencies supportive of ESD (systems thinking, futures thinking, normative thinking, action-oriented thinking, interpersonal skills, and integrative meta-competences). We report below based on analyzing 52 relevant comments: systems thinking (11 comments), futures thinking (7), normative thinking (9), action-oriented thinking (7), interpersonal skills (8), and integrative meta-competences (9).

SCALAR-SHIFTING PROCESSES AROUND SYSTEMS THINKING

The first competencies related to systems thinking, particularly creating a constructive understanding between the general level of the SDGs and the ways they presented in particular locations. One respondent wrote "I learned a lot about analysing global issues and breaking them down." A key learning element was in dealing with the shifting in focusing on a specific challenge, and the revelation that simple cause–effect relationships were part of a tapestry of other factors, all needing inclusion in the analysis.

> For the solution we had to take into account the culture, the healthcare system of Indonesia, the different stakeholders who are involved in this healthcare system and who are involved in the recommendation but also the legal system, in our project the ministry of health was a very important factor to take into account. … In my opinion our recommendation is applicable on a small scale in Bali, but also in other cultures and on a larger scale.

Likewise, students realized experientially that sometimes trivial factors were potentially critical for delivering solutions: "good learning conditions" for education projects included elements such as "air conditioning" as well as motivated teachers, relevant learning materials, and good equipment. This had consequences for how students created general explanations of solutions to and more general recommendations for, the SDGs:

> I found that when you try to look deeper into one specific grand challenge, it becomes clear that there is not just one big perfect solution for the challenges.

The scalar shifting in the systems thinking competence involved breaking things down and building them back up, driven by requirements to undertake a separate research project to the overall placement. The scalar shifting here appears driven by coupling of the research project with the placement, with teachers' and supervisors' input critical in encouraging students to use their lived experiences during placements to inform their diagnoses and interpretations.

SCALAR-SHIFTING PROCESSES AROUND FUTURES THINKING

The second competency was futures thinking, where students learn an attitude of creating a better future by identifying mechanisms, pathways, and tools to build those more desirable futures. Before the placements, a number of groups used a naïve knowledge gap models in their posters, arguing that people took bad choices because they lacked knowledge, making a suitable solution an information campaign. One scalar shift was realizing that information campaigns were inadequate for shifting deep-seated mind-sets. As one volunteer at a wildlife project wrote:

> Doing an internship at this organisation has shown me how hard it is to change the mindset of almost an entire country, despite sustainability and nature conservation being such important topics.

An additional element was realizing that real problems did not correspond perfectly to a single SDGs, and solving real problems might involve addressing one or more SDG. A student working in an education project noted that a key element of the project's success was providing pupils with free snacks, but a better solution would be better food security in those pupils' households. There was also an appreciation that although projects were always worthy, many were insufficient to deliver desirable futures. One report noted:

> [The Host Organisation] seem to only focus of fighting the symptoms and not the causes of the problem, which is in the end probably less effective. But as we learnt, it is difficult to do effective work when you do not have the materials to begin with.

We infer here that part of the learned competency involves not just criticizing the organization (symptom-focusing) but also developing recommendations to address the problem, namely better resourcing that organization. Part of futures thinking involves a continual shifting between the two scales, of theory and practice, using theory creatively to identify why practices might be inadequate to address these problems.

SCALAR-SHIFTING PROCESSES AROUND NORMATIVE THINKING

Normative thinking involves students' perspectives that particular situations are unacceptable and demand action. The UT has a reputation for being primarily focused on producing technological solutions (Cinar, 2019); student activism primarily involves becoming a student or study associations' board member, not political activism. Against this background, it was surprising in some reflection reports to read students taking strongly normative positions regarding the moral unacceptability of the conditions they had encountered during their placements. One student reported "finding my passions and trust in myself," another student spoke of the struggles that people faced in trying to live decent lives, and a third student spoke of that they wanted "other people to experience this as well so that we are able to solve some of the world's problems."

The change involved in acquiring this competency was neatly encapsulated by the student writing "I have become more of a global citizen in this world" highlighting two key elements. First, is the idea of global citizenship, which is sharing common entitlements to and responsibilities for other communities. One student noted that they had learned to feel their own responsibility for the problems:

> Learning about these challenges, gave me on the one hand an anxious feeling that we as humans are able and willing to pollute a whole planet. However, it also really opened my eyes that I am contributing to it as much everybody else does. Therefore, I want to change my behavior and live more environmentally friendly.

Second, is the idea of "becoming," which the learning process involved practical experiences convincing individuals that their general judgement was correct, as one student wrote:

> In this case, seeing people that were undernourished and children that had not eaten the previous day really opened my eyes to the real problems in the world.

Thus, diagnosing injustice was just abstract, stating it was not fair that people were dying of a lack of access to water. Students experienced that injustice practically, trying to address a symptom and finding themselves also trapped in vicious cycles, hindered producing solutions. This helped students to realize that social as well as technological changes were necessary.

SCALAR-SHIFTING PROCESSES AROUND ACTION-ORIENTED THINKING

What we term action-oriented thinking is the competence of using knowledge to achieve ends, without allowing knowledge creation process' demands to inhibit the delivery of the eventual outcome. As Wiek et al. (2011) describes it:

> In simple terms, this competence is about being able to 'get things done'. This involves familiarity with real-world situations and relationships, political understanding, challenging positions at the right time, being able to solve logistical problems, using language that non-academics are comfortable with, working with deadlines that governments insist on, and so forth (p. 210).

The students learned the attuning of externally created knowledge to ensure that knowledge could be usefully deployed in these local contexts. One good example, already alluded to, was students began with a naïve knowledge gap model, while experience demonstrated that was not the case. As one student noted:

> In [region], public awareness of environmental problems is quite high. For instance, everybody knows the fish stock is declining. [The host organisation] is pushing projects in order to restrict foreign fishing and stop poaching, but nobody would ever think of trying to eat less fish. Since I am a vegetarian since many years, the topic came up quite often and the reaction was always the same. The idea of trying to eat less fish always caused confusion and apprehension.

One learning effect came through a realization that the scalar mismatch that students were learning also had consequences for individuals' own behavior. This helped raising awareness not just of the problem, but how individual choices cumulate to undermine solutions. One student wrote:

> This makes me feel like the Grand Challenge of Sustainability might be one of the biggest challenges of all, since it needs a change of mind and behaviour in three generations, over the whole world. People got used too much to their modern consumerism and might not want to restrict themselves for an abstract, higher goal.

Projects' role was important here, in bringing home that these grand challenges were made everywhere, thereby creating an understanding that individual or group action could address them. Students wrote that it was the experience of "seeing with their own eyes" that made them realize the solution lay in modifying their own behavior and not just assisting local communities.

SCALAR-SHIFTING PROCESSES AROUND INTERPERSONAL SKILLS

The interpersonal competence relates to the capacity to connect with others to create and transfer actionable knowledge contributing to the constructive solution of problems constitutive of SDGs. This involved teachers and students building connections permitting knowledge flow between different actors, contributing to the various pedagogic steps. An important element of the scalar shifting involved here was in the working in multidisciplinary teams on academic questions and then working in multi-stakeholder teams in the volunteering activities, and combining those two constructively in the final report. One student said that the collaborations prior to the placement demonstrated to them the value of learning about working with other disciplinary backgrounds, as well as the differences in approaches that emerge from the different ways that disciplines frame problems and solutions:

> One of the very useful aspects of the minor was the collaboration with students from other disciplines. I learned a lot from working with people who might tackle problems on a different way than I should do it.

While another student noted that the group work on theory-stimulated changes in the orientation of that student toward agency and responsibility (see normative thinking):

> It was also very helpful to have the group work regarding the grand challenges. In this element I learned a lot about the grand challenges themselves of course, but also about how to start thinking about making a change. It was good to see how things do and do not work and it gave me a feeling that every person could make a difference to solve a big problem or even make a small contribution.

It can be common to think of stakeholder partnerships from a constructive perspective in which everyone is committed to delivering a common set of goals. However, another important set of skills was in learning to work with what might be considered as "unruly beneficiaries," that do not necessarily share the same view of the value of an activity. One student reflected:

> I learned to look for opportunities and take action instead of waiting passively. Some situations asked action without waiting. Especially with the young children. If I was not careful at a moment, they were gone or they argued.

A final element of the scalar-shifting process around interpersonal skills process was in constituting the research subjects as having an agency and knowledge within the process, and identifying ways in which their knowledge could be accessed and used to create a wider global story. One student noted that they[3] had initially been opposed to undertaking interviews, but rather than simply finding them useful, they had actually found them stimulating and indeed a useful source of knowledge that could be fed back into the overall process.

> For the social aspects of my research I had to do interviews, which I had never done before. To be honest, at first I did not really want to interview people and tried to avoid it. But after I got the results back, I was quite interested and surprised what people answered. They came up with *more new ideas than I expected* (authors' own emphasis).

SCALAR-SHIFTING PROCESSES AROUND INTEGRATIVE META-COMPETENCES

Integrative meta-competences are capacities to organize activities to ensure that the five other competencies are coupled to change activities. Part of coupling was built into the course organization, with students being forced to follow a set of protocols that sought to stimulate them to take those scalar shifts. The minor provided students tools with which to achieve this integration, and the evaluation framework incentivized students to deliver it. Conversely, it was left to students working on their projects as to how they would pull those elements together in practice to achieve the overall change. As one student noted:

> [At the start of the minor] I didn't really have an idea of how does it look like to implement new innovations and to solve the challenges. My preparation was mostly theoretical, and I got the real insight into these topics when I came to the host organisation. Here, I saw what it really means to work sustainable and to do something good for future generations.

It is worth stressing the level of competency being developed around integrative meta-competencies was relatively limited. Students could perform well simply by being aware of the challenges that existed in trying to deliver an integrated solution. One student wrote:

> In my case, I have the perception that I was able to form an impression of the obstacles and main problems of the grand challenge like inequality and non-availability of education for all. Moreover, I did not only encountered problems but also came in touch with solutions.

DISCUSSION AND CONCLUSIONS

In this chapter, we have sought to address the research question of whether learning communities can be constructed spanning from "classroom to local project" thereby coupling practical experiences to theoretical learning outcomes. We addressed using a model of learning competencies, where learning competencies are demonstrated through "scalar shifting": theoretical knowledge drives practical action, and practically derived knowledge informs a wider theoretical contribution. It is important to stress that we are able to demonstrate that there are projects that complete the full cycle – at least one student has been able to publish their research findings as an academic article, demonstrating that they have made a conceptual diagnosis, undertaken a practical placement, and used it to contribute conceptually to debates around SDGs (in this case, around SDG17, Capacity Building, cf. Cunha, Mensing, & Benneworth, 2018).

Our contention is that "coupling" between the theory-centered and practical-centered elements of ESD emerges through pedagogic activities building competencies involving a scalar shift between the two scales of learning experience. This coupling process involves four actors, all of which ensure that the student is able to embody the learning competencies. First, and most obviously is the student, who both develops a normative sense that the world should be different to solve the obvious injustice that lies behind the SDG, while integrating other competencies to produce the full cycle of transformative change. Second, is the local supervisor, who assists with diagnosing and building the necessary systems thinking, while actively participating in the learning community that the learner constructs around themselves spanning classroom and placement. Third, are placement partners, who exert agency through the project, with the project becoming an instrument for their agency exertion. Finally, is the wider policy and academic community providing a conduit by which the locally specific learning can inform understandings and debates around the SDGs.

We therefore argue that effectively delivering ESD involves not simply a process of teaching students, but of situating students within these wider learning communities. This gives learning a dualistic character: first, learners undergo a learning journey where they achieve change and learn to shape that change in a specific local context, while coupled to wider academic and policy debates. This provides a means to nuance our understanding of the diagnosis we made earlier in the chapter, namely that the problem is not specifically a "fairness" challenge ensuring that students are only assessed on the basis of what they can reasonably achieve. The challenge is of constructing the necessary learning communities where this coupling can take place, linking classrooms with sites of change, linking teachers and community activists allowing students to undergo these revelatory learning experiences. This implies that the more general challenge for ESD in higher education is finding ways to institutionalize these border-crossing learning communities, allowing them to steer learners toward producing these competencies more systematically.

We are cautious in making sweeping generalizations from a relatively small retrospective study from a privileged situation built on substantive institutional efforts. We are nevertheless struck by how this close coupling appears necessary to confront students with the SDGs' urgency and immediacy in pedagogic frameworks that allow them to develop this more abstract appreciation that these are not exclusively local problems. This mechanism – scalar shifting – appears to have lessons for higher education – and education more generally – in building students that have an awareness of and a capacity to address local manifestations of these wider problems. More consideration is needed of this mechanism to truly realize ESD's capacity to actively contribute to empowering diverse communities to build a better world.

ACKNOWLEDGMENTS

We would like to thank the four cohorts (at the time of writing) of the Crossing Borders module who have provided the basis for the insights reported in this chapter. We also would like to thank the Student Mobility Office for providing us with data regarding student numbers and bursary funding provided for placements. Many thanks are also due to two anonymous reviewers for their comments on a first draft of the text. All errors and omissions remain the responsibility of the authors.

NOTES

1. These statements and examples presented in this section were compiled from tutorial discussions with students doing preparatory work within the teaching activity described in this chapter, antecedent to preparing the SDG poster (*qv*). More information on that activity can be found in the second half of the Methodology section.
2. This example was developed to explain to course students how the different elements/activities were integrated.
3. We here follow the common practice among native English editors of using "they" as a singular pronoun where it is not germane what the gender of the student is.

REFERENCES

Benneworth, P., & Osborne, M. (2016). Institutional approaches to supporting university entrepreneurship education: fitting entrepreneurship within university institutional architectures. *Journal of the European Higher Education Area, 2*, 1–26.

Cinar, R. (2019). Delving into social entrepreneurship in universities: Is it legitimate yet? Regional studies. *Regional Science*, *6*(1), 217–232. doi:10.1080/21681376.2019.1583602

Cunha, J., Mensing, R., & Benneworth, P. (2018). A conceptual framework to understand student volunteerism. *Tertiary Education and Management* (Online first), 1–14. doi:10.1007/s11233-018-09012-x

Dale, A., & Newman, L. (2005). Sustainable development, education and literacy. *International Journal of Sustainability in Higher Education*, *6*(4), 351–362.

Fayolle, A., Verzat, C., & Wapshott, R. (2016). In quest of legitimacy: The theoretical and methodological foundations of entrepreneurship education research. *International Small Business Journal*, *34*(7), 895–904.

Juliani, D., Tamilis da Silva, A., Cunha, J., & Benneworth, P. (2017). Universities' contributions to sustainable development's social challenge: A case study of a social innovation practice.

International Journal of Social Ecology and Sustainable Development, *8*(3), 1–18. doi:10.4018/IJSESD.2017070101

Leicht, A., Heiss, J., & Byun, W. J. (2018). *Issues and trends in education for sustainable development* (Vol. 5). Paris, France: UNESCO Publishing.

Molderez, I., & Fonseca, E. (2018). The efficacy of real-world experiences and service learning for fostering competences for sustainable development in higher education. *Journal of Cleaner Production*, *172*, 4397–4410.

Quality Assurance Agency (QAA). (2014). *Education for sustainable development: Guidance for UK higher education providers*. Gloucester: Quality Assurance Agency for Higher Education.

Shephard, K., Rieckmann, M., & Barth, M. (2018). Seeking sustainability competence and capability in the ESD and HESD literature: An international philosophical hermeneutic analysis. *Environmental Education Research*, 1–16. doi:10.1080/13504622.2018.1490947

United Nations Educational, Scientific and Cultural Organization (UNESCO). (2009, March 31–April 2). Bonn declaration. In *Final communique from the UNESCO world conference on education for sustainable development*, Bonn, Germany.

United Nations Educational, Scientific and Cultural Organization (UNESCO). (2012). *Exploring sustainable development: A multiple-perspective approach*. Paris, France: Education for Sustainable Development Section (ED/PSD/ESD).

United Nations Educational, Scientific and Cultural Organization (UNESCO). (2019). Sustainable development goal 2: End hunger. Sustainable Development Goals Knowledge Platform. Retrieved from https://sustainabledevelopment.un.org/sdg2. Accessed on April 15, 2019.

Wiek, A., Bernstein, M. J., Foley, R. W., Cohen, M., Forrest, N., Kuzdas, C., … Withycombe Keeler, L. (2016). Operationalising competencies in higher education for sustainable development. In M. Barth, G. Michelsen, I. Thomas, & M. Rieckmann (Eds.), *Routledge handbook of higher education for sustainable development* (pp. 241–260). London: Routledge.

Wiek, A., Withycombe, L., & Redman, C. L. (2011). Key competencies in sustainability: A reference framework for academic program development. *Sustainability Science*, *6*(2), 203–218. doi:10.1007/s11625-011-0132-6

ABOUT THE CONTRIBUTORS

Luis Kluwe de Aguiar is the Course Manager for B.Sc. (Hons) Food Degree Programs and Senior Lecturer on Food Security and Sustainability as well as Food Marketing. Prior to becoming a full time academic, he worked in different agribusiness-related roles, namely as Agricultural Attaché at the Foreign and Commonwealth Office British Embassy, Brazil, the Ministry of Foreign Relations of Japan; Commercial Manager of agricultural storage systems; Food Technologist for barley malting, and Sales Promoter for an importer of horticultural seeds. He has done international consultancy work and has an extensive publication portfolio in the field of food marketing and sustainability.

Meltem Alemdar is Associate Director and Senior Research Scientist at Georgia Tech's Center for Education Integrating Science, Mathematics, and Computing. Dr Alemdar has experience evaluating programs that fall under the umbrella of educational evaluation, including K-12 educational curricula, K-12 STEM programs after-school programs, and comprehensive school reform initiatives. Across these evaluations, she has used a variety of evaluation methods, ranging from a multi-level evaluation plan designed to assess program impact to methods such as program monitoring designed to facilitate program improvement. She received her Ph.D. in Research, Measurement, and Statistics from the Department of Education Policy at Georgia State University.

Anis Azizi is a two-time scholarship recipient from the Public Service Department of Malaysia and an international student who recently graduated with a Bachelor of Science in Psychology from the University of Auckland. She was a participant in the university's science scholars program and conducted a research project examining the meaning of clothes to young people under the supervision of Dr Niki Harré. Anis plans to further her engagement with initiatives that contribute toward sustainability in Malaysia by joining a fabric recycling movement and installing fabric recycling bins in her neighborhood with a local organization named Kloth Malaysia.

Patrick Blessinger, Ed.D., is an Adjunct Associate Professor of Education at St. John's University, a Math and Science Teacher with the New York State Education Department, and Chief Research Scientist of the International Higher Education Teaching and Learning Association (in consultative status with the United Nations). Blessinger is the editor and author of many books and articles and he is an Educational Policy Analyst and contributing writer with UNESCO's Inclusive Policy Lab, *University World News*, *The Hechinger Report*, *The Guardian*, and *Higher Education Tomorrow*, among others. Blessinger teaches courses in education, leadership, and research methods and he serves on doctoral dissertation committees. Blessinger founded and leads a global network of educators focused on teaching and learning and he is an expert in inclusion, equity, leadership, policy, democracy, human rights, and sustainable development. Blessinger provides professional development workshops to teachers and professors and regularly gives presentations and keynote addresses at academic conferences around the world. Blessinger has received several educational awards, including: Fulbright Senior Scholar to Denmark (Department of State, USA), Governor's Teaching Fellow (Institute of Higher Education, University of Georgia, USA), and Certified Educator (National Geographic Society, USA).

Paul Benneworth is a Professor of Innovation and Regional Development at the Department of Business Administration and Social Sciences at the Western Norway University of Applied Sciences and a Senior Researcher at the Center for Higher Education Policy Studies, University of Twente, the Netherlands. Paul's research focuses on the dynamics of innovation and societal change and in particular the ways in which partnerships of researchers, policy-makers, and practitioners try to use various kinds of knowledge to stimulate socioeconomic development. Paul leads the Working Group on SSH Research Impact in ENRESSH, the COST-funded European Network for Research Evaluation in the Social Sciences and Humanities. He has been Chief Scientist of projects for a range of research funders including the European Framework Program/ERA-NET, research councils in the UK, the Netherlands and Norway as well as a wide range of government and not-for-profit funders. He is the editor and author of eight scientific books, including *The Impacts and Future of Arts and Humanities Research* (Palgrave, 2016) and *Universities and Regional Economic Development* (Routledge, 2018). He has published more than 30 peer-reviewed scientific articles in the last decade, serves as reviewer for a range of journals, research councils, and funding agencies internationally and is currently acting as special scientific adviser to three research council funded projects in Spain, Austria, and Portugal.

Penny Brothers is a New Zealander, educated at the University of Auckland and Stanford University. She was in the School of Chemical Sciences at the University of Auckland from 1988 to February 2019 when she took up a position as Director of the Research School of Chemistry at Australian National University. She has been a Visiting Professor at the Universities of California at Davis and at Berkeley, the Heidelberg, Münster, Burgundy, Peking University, the Arctic University of Norway and Los Alamos National Laboratory (as a Fulbright Senior Scholar). Professor Brothers is a Principal Investigator in the MacDiarmid Institute for Advanced Nanomaterials and an Associate Editor of *Chemical Communications*. She is currently investigating the chemistry of boron coordinated to porphyrin and corrole ligands, BODIPY fluorophores for sugar recognition, and surface patterning using molecular pentagons. While at the University of Auckland she taught Chemistry at all levels from introductory to postgraduate, including Green Chemistry and chemical principles applied to global sustainability issues.

Alice Cassidy, Principal at In View Educational Development, is a wildlife ecologist, science educator and educational developer with degrees from the University of Victoria, McGill University and the University of British Columbia (UBC). She has conducted environmental impact studies and led bird identification. At UBC she taught upper level biology courses and coordinated and designed first year courses focused on scholarly writing. When Associate Director of the UBC's campus-wide teaching support centre, and now as a consultant, she designs professional development for teachers at all levels, both face to face and online, and directed large-scale educational programs. She works as an editor and reviewer, recently for Elsevier Science. She has published and presented on many aspects of science, sustainability, scholarship of teaching and learning and educational development, including a recent book, *Learning from Conferences* (Brill). Sustainability projects, funded by the Educational Developers Caucus, include a sustainability repository and a set of recorded interviews of those working in sustainability educational development, both freely available online https://cassidyinview.wordpress.com/ along with many other resources for educators. She often works with students as co-scholars. She focuses on the importance of sustainability and increased awareness of the natural world, leading to conservation.

Ties Coomber is a Ph.D. candidate under the supervision of Dr Niki Harré and Dr Quentin Atkinson at the University of Auckland. His undergraduate background is in Anthropology, Philosophy, and Psychology. His Ph.D. research focuses on "oneness" concepts, which measure perceptions of being profoundly connected to other, often larger entities, like social groups, nature, or even all of existence. In particular, the research focuses on how these many and varied concepts can be arranged into an overarching typology, and how different dimensions of this typology relate to pro-social and pro-environmental values, motivations, and behaviors.

Kata Dosa is a Postdoctoral Scholar in the Center for Teaching and Learning. She earned her Ph.D. in Environment and Resources from the University of Wisconsin-Madison, where Dosa conducted discipline-based education research looking at environmental decision-making and reasoning processes. During this time, she also completed the DELTA Certificate for Research, Teaching, and Learning. Dosa holds a Master's degree in Environmental Science and Environmental Biology from Eotvos Lorand University. She is currently working on developing workshops for faculty and graduate students, and supports future faculty and teaching assistant development programs. Dosa's current research interests are teaching-as-research, incorporating sustainability across the curriculum, team science, and competency development in higher education.

Laura Franco Garcia is Assistant Professor at CSTM Governance and Technology for Sustainability Department of the University of Twente. Franco-García holds an Environmental Engineer's degree from Metropolitan Autonomous University in Mexico City, a Master's degree on "Management and Treatment of Solid Wastes" (Institute Nationale des Sciences Appliquées in Lyon, France), a Ph.D. on "Environmental Chemistry" (Université Claude Bernard Lyon I, France), a Pos-doctorate study related to "Catalytic and Nano Sciences" (Hokkaido University, Japan), and research-stay at the Ashridge Business School, Great

Britain. Franco-García lectured and researched at National Autonomous University of Mexico, Metropolitan Autonomous University, and Tecnológico de Monterrey. These represent some of the most recognized high educational institutions in Mexico. In 2003, she became Director of the first Master's program on "Sustainable Development Sciences" and she was head of the "Sustainable Development" Department at Tecnológico de Monterrey, Campus Mexico State, till 2006. Currently, she teaches in the Masters of Energy and Environmental Management and the program (at bachelor level): "Crossing borders, international grand challenges and strategies for technology, sustainability and development." Franco-García is one of the European coordinators of the global network of "Greening of Industry Network (GIN)", www.greeningofindustry.org. GIN organizes conferences in three regions: Asia, America, and Europa. The latest one (October 2019) was dedicated to topics associated with "Environment and Governance, Innovations towards Sustainability".

Alana Griffith is a Lecturer in Sociology specializing in Social Policy at the University of the West Indies, Cave Hill Campus (UWI). She holds a Ph.D. in Social Policy from the University of Kent (UK). She began lecturing part-time with UWI in 2004 while working in the Prime Minister's Office Barbados as a Research Officer. She joined the Faculty as a Full-time Member later in 2005 and was awarded the Best Female Lecturer in the Faculty of Social Sciences for academic year 2005–2006 by the Social Sciences Society. She has published a number of scholarly articles, among which is "The Feasibility of Flipping: An Exploratory Analysis of the Flipped Class in Caribbean Sociology" in *Journal of Effective Teaching* in 2017. In June 2018, she received a UWI Guild of Students/ Centre for Excellence in Teaching & Learning (CETL) Award for Outstanding Contribution to Student Learning. She holds a Postgraduate Certificate in University Teaching and Learning and is a Member of the University of the West Indies' Teaching and Learning Committee and the Faculty of Social Sciences' Teaching and Learning Committee. She is the recipient of the Principal's Award for Excellence in Outstanding Teaching for 2018–2019.

Dawne Gurbutt is an experienced Practitioner and Academic with a background in Public Health. After qualifying as a nurse, midwife, and health visitor, she worked in areas of social deprivation across the North of England working directly with families and communities. As an experienced academic she holds qualifications in a range of discipline areas including health, education, literature, and social sciences and has managed a wide range of provision in different HE settings. This experience has led to a keen interest in collaborative practice and education. Dawne has held strategic roles including working across the four nations of the UK as Discipline Lead for Health in the Higher Education Academy, UK. Her role in the HEA included commissioning, designing, managing, and delivering projects and programs on pedagogy; public involvement; and collaborative learning, sustainability, and creativity in teaching and learning; she led on national initiatives including the development of an integrated PPI strategy for Health, internationalization of the health curriculum, and developing/commissioning resources to support HEI colleagues in Health. She represented the HEA on national forums on simulation in HE and creativity in the curriculum and Flipped Classroom approaches. She leads on Collaborative Learning and Teaching at UCLan. She is committed to working with stakeholders and service users in IPE to improve patient outcomes. She is an Honorary Fellow of the Centre for the Advancement of Interprofessional Education (CAIPE) and is also a member of the CAIPE Board where she co-leads the Learning and Teaching group. Working in close collaboration with individuals and groups has been a continuous and consistent thread throughout her career in health and education.

Russell Gurbutt is an inspirational and innovative Educator working in Health and Social Care with a focus on decision making in relation to leadership and management. Following a successful military career in the Royal Navy which included being awarded the ceremonial sword following graduation from BRNC Dartmouth he saw active service in the Falklands war. The experience of the Royal Navy provided leadership and management experience across a range of scenarios which was later developed in Healthcare and Educational management settings. He has extensive experience of the third sector having worked in a leadership role in quality and development for a National Charity, a Healthcare consultancy, and served as a Management Trustee. Committed to sustainable and innovative education, he has a proven track record of agility in developing curricula, engaging students, supporting staff, and enabling learning communities with regard to online learning and teaching. He has worked in the UK and abroad.

Euan Hague, Ph.D. is a Professor of Geography and Director of the School of Public Service at DePaul University. An Urban and Cultural Geographer, Dr Hague's work has appeared in peer-reviewed journals including *Annals of the American Association of Geographers*, *Professional Geographer*, and *Cultural Geographies*. His recent analyses exploring housing in Chicago focuses on gentrification and community resistance to processes of displacement. Dr Hague's books include *Regional and Local Economic Development* (Palgrave Macmillan 2011, co-authored with Cliff Hague and Carrie Breitbach) and *Neoliberal Chicago* (University of Illinois, 2017, co-edited with Larry Bennett and Roberta Garner). He is one of the founders of DePaul's sustainable urban development graduate degree program and received the Coalition of Urban and Metropolitan Universities' 2019 Barbara A. Holland Scholar-Administrator Award.

Niki Harré is a Professor in the School of Psychology at the University of Auckland. Her recent research projects have focused on sustainable communities and schools, positive youth development, and political activism. She teaches social and community psychology with a particular emphasis on the psychology of sustainability. Niki is the Coordinator of a three course interdisciplinary teaching module on sustainability available to B.A. and B.Sc. students at the University of Auckland. In 2007, Niki edited, with Quentin Atkinson, the book *Carbon Neutral by 2020: How New Zealanders Can Tackle Climate Change* (Craig Potton). Two of her most recent two books were published by Auckland University Press in 2018: *The Infinite Game: How to Live Well Together* and *Psychology for a Better World: Working with People to Save the Planet*.

Jennifer Hirsch is an Applied Anthropologist specializing in sustainability, cultural diversity, collaborative governance, community development, networks, and experiential education. She is recognized nationally for fostering grassroots participation in sustainability and climate action, using methods from visual and environmental anthropology, popular education, asset-based community development, participatory facilitation, and museum education. Since August 2015, she has been the inaugural Director of the Center for Serve-Learn-Sustain at the Georgia Institute of Technology. Serve-Learn-Sustain is a campus-wide academic initiative launched in January 2016 to prepare students to use their disciplinary expertise to address the world's most pressing sustainability challenges in ways that advance community well-being and create thriving natural systems. Dr Hirsch is Adjunct Associate Professor in the School of City and Regional Planning at Georgia Tech and serves on the faculty of The Asset-Based Community Development Institute hosted by DePaul University.

Robert Huddleston is currently a Research Degree Supervisor at Birmingham City University, Faculty of Arts Design and Media, specializing in art and design education. With a Master's degree in Fine Art from the Royal College of Art, London, he has held senior lecturing positions at Winchester School of Art, University of Southampton, (1998–2010) and Birmingham Institute of Art and Design (1981–1998), and has taught as a Visiting Lecturer at a number of art and design schools in the UK and at Tokyo Zokei University, Japan. His academic experience includes consultancy on programs of international education cooperation and appointments as external advisor on art and design curriculum development. His practice has included design for major European textile companies and a photography project with the photographer and filmmaker, Dan Brackenbury. This project documents the changing urban spaces of the port city of Dalian in China for the archives of the Royal Institute of British Architects (London). It was undertaken between 2012 and 2014. During this period, Robert was leading two Winchester School of Art Bachelor's programs, taught at Dalian Polytechnic University, in China.

Ben Jongbloed is a Senior Research Associate at the Center for Higher Education Policy Studies of the University of Twente, the Netherlands, where his main research themes are the governance and financing of higher education. As part of this, he is also working on topics such as performance indicators and entrepreneurship in higher education (in the U-Multirank, respectively the HEInnovate project). As a teacher, tutor, and coordinator, he is involved in the design and delivery of the University of Twente's minor program Crossing Borders.

Ellinor King has a Bachelor of Arts with Honors and a Master of Arts in Psychology. She graduated in 2018, with a thesis that focused on utilizing higher education and interdisciplinary teaching to promote sustainable and ethical consumption in university students. It was a part of a larger clothing study carried out by the University of Auckland which was supervised by Niki Harré. Ellinor is interested in continuing to explore this field and using the power of education to strive toward a more sustainable society.

Renze Kolster is a Research Associate at the Center for Higher Education Policy Studies (CHEPS) of the University of Twente, the Netherlands. He received a Master degree in Public Administration from the University of Twente and a M.Phil. degree from the Erasmus Mundus program "European Master in Higher Education" (inter alia, granted by the University of Oslo). At CHEPS, Renze mainly researches topics related to internationalisation of higher education, higher education relevance, excellence in higher education, and graduate employability. These are also the topics of his forthcoming Ph.D. dissertation. As a Visiting Researcher, Renze is connected to the University of São Paulo in Brazil. As a teacher, tutor, and coordinator, he is involved in the design and delivery of the minor Crossing Borders.

Martha Elena Núñez López has been a Professor at Tecnologico de Monterrey in Mexico since 2003, and she is currently the Dean of the Architecture Department for the South Region (June 2017–present). She is pursuing her Ph.D. at Birmingham city University, UK (September 2016–present). Her research is about: "Supporting Mexico's Sustainable Development through Exploring Innovative Approaches to the Product Design Curriculum in Higher Education." She holds two Master degrees: Industrial Design and Product Innovation (Graduated with the Highest Honors) and Construction Management and Engineering (Excellence Scholarship). She holds a Bachelor degree in Architecture (Graduated with the Highest Honors). Other positions that she has occupied at Tec de Monterrey are: National Coordinator of the Architecture, Design and Arts Academies (May 2016–August 2017), Academic Director of the Industrial Design Department (January 2011–January 2016), and Director of the Industrial Design Bachelor (January 2008–January 2011). Her passion is innovating in teaching and learning; she has been shortlisted three times at the QS-Stars Reimagine Education Awards, USA (2015–2017) with her projects: Remote teaching for student interaction with foreign professors and academic leaders, Use of Innovative Learning Spaces, and Real-time remote classes. She is also a judge for the Reimagine Education Awards (2016–present). In May 2017, she won the "Inspiring Professor Award" at Tec de Monterrey. She has organized successful Summer Programs at UK, Japan, China, and Thailand. She has invited foreign visitor professors (a one-year program in Mexico) from Poland, Japan, and Brazil; as well as Academic Leaders from India, UK, Japan, Brazil, Netherlands, Poland, and Puerto Rico. She has presented her Research and work at different forums: ASME 2018 International Mechanical Engineering Congress and Exposition, Pittsburgh, USA, The Festival of Teaching at Birmingham City University, 2017, and /Art Gallery in Kanazawa, Japan, 2016.

Roberto Pablo Martínez Lozano is a Professor at Tecnológico de Monterrey in México since 1999, where currently he is the Program Director of the Multicultural Program (2014–present). He holds a Ph.D. in Educational Innovation, has a Master in Administration and Information Technology, and holds a Bachelor degree in Engineering and Electronic Systems. Other positions that he has occupied at the Tec de Monterrey are: Professor at Computer Science and Networks 1999–2004, Coordinator of Math Department 2004–2005, Director of the Bilingual Program 2005–2007, and Director of the Bicultural and International Program 2007–2014. For him to be a professor is a blessing and a privilege for doing what he most desires: "To impact others by leaving a footprint on the lives of his students." He has written several articles about the use of the Hearth Math technique to reduce anxiety and stress when faced with math tests. He won the Teacher Excellence Award in May 2013. He has given several conferences at the International Congress of Educational Innovation in 2015, 2016, and 2018.

Louise Manning, Professor in Agri-food and Supply Chain Security, has worked for over 30 years in the agri-food supply chain in a range of roles. She lives on a mixed-enterprise family farm in Herefordshire. Her expertise is in the area of food integrity including food safety, food quality, food crime, and governance, social, and corporate responsibility. She advises both government and corporate businesses on supply chain risk assessment techniques and associated mitigation strategies. Louise has published in a range of peer-reviewed journals and regularly presents at conferences and writes for trade press and electronic media.

Andrea Michelle Mead is a Professional Teaching Fellow in the School of Psychology at the University of Auckland. Andrea teaches in and coordinates a large stage one psychology course. Andrea is an active member of the Faculty of Science Sustainability network which led to the development of an interdisciplinary/cross-stage sustainability teaching case study.

Winston Moore is presently Professor of Economics and Director of Graduate Studies and Research at The University of the West Indies, Cave Hill Campus. Prior to this, he held the position of Senior Economist at the Central Bank of Barbados. Professor Moore is a very active public commentator on issues related to national development and is a former President of the Barbados Economics Society. His recent research has examined the issues surrounding the green economy, private sector development, as well as the economic impact of climate change on tourism. Professor Moore has published more than 90 peer-reviewed papers, books, and book chapters. He holds a Ph.D. in Economics from the University of Surrey, a M.Sc. in Economics from the University of Warwick, and a B.Sc. in Economics from the University of West Indies, Cave Hill. He also holds a Postgraduate Certificate in University Teaching and Learning and is a member of the University of the West Indies' Teaching and Learning Committee.

Michael W. Mulnix, Ph.D. is Founding President of the American University of Iraq – Baghdad. Prior to that he was also Founding President of the American University of Kurdistan. He has lived in the Middle East for the past nine years, serving as an international educational consultant in Saudi Arabia, Kuwait, and the UAE. He has also consulted extensively in Mexico. Dr Mulnix served as Provost/President of Antioch University Santa Barbara, as Professor of

Marketing and Dean of Graduate Studies and Research at the University of the Incarnate Word in San Antonio, Texas, and as Vice President for Institutional Advancement at Our Lady of the Lake University, also in San Antonio. He was Executive Director of University Relations and in charge of the Office of Governmental Relations at the University of Nebraska – Lincoln for nine years and Director of Public Relations at the American Association of State Colleges and Universities in Washington, D.C. He also served as Executive Director of University and Government Relations and Executive Assistant to the Chancellor at the University of Alaska in Juneau. Dr Mulnix was trained as a journalist and was an award-winning reporter and columnist for several newspapers in Colorado and Alaska.

Sunni Newton is currently a Research Associate at the Georgia Institute of Technology in the Center for Education Integrating Science, Mathematics, and Computing. Her research focuses on assessing the implementation and outcomes of educational interventions at the K-12 and collegiate levels. She received her M.S. and Ph.D. in Industrial/Organizational Psychology from Georgia Tech.

Sarah Nyrose is a Naturopathic Physician practicing in Victoria, British Columbia. She completed naturopathic medical school at the Boucher Institute of Naturopathic Medicine and holds an undergraduate degree in Environmental Sciences from the University of British Columbia. Sarah is passionate about medicine, nutrition, and sustainability.

Raghu Pucha is a Senior Lecturer at the Woodruff School of Mechanical Engineering, Georgia Institute of Technology, in the area of CAD/CAE and Manufacturing. Dr Pucha teaches CAD, CAE, and mechanics courses at Georgia Tech., and conducts research in the area of developing computational tools for the design, analysis, and manufacturing of advanced materials and systems. Dr Pucha has three provisional US patents and co-authored over 65 research papers. He is honored with Undergraduate Educator Award in 2012 and Geoffrey G. Eichholz Faculty Teaching Award in 2015 from the Center for Enhancement of Teaching and Learning at Georgia Tech. Dr Pucha's current class-room research interests are learning-centered instruction with context and sociotechnical project-based learning in engineering courses.

Howard Rosing is the Executive Director of the Steans Center at DePaul University. The Center offers university-wide support to faculty and students through developing collaborative teaching, research, and training partnerships with community organizations throughout Chicago and internationally. The center also houses the Egan Office for Urban Education and Community Partnerships and the Asset-Based Community Development Institute. Dr Rosing serves as a Faculty Member in Community Service Studies and an Affiliate Faculty Member in Geography, Sustainable Urban Development (MASUD), and Community Psychology; he is a Co-founder of and co-directs MASUD. Dr Rosing is a Cultural Anthropologist whose research focuses on sustainable urban food systems and food justice movements in Chicago and the Dominican Republic.

Sarah Saeckel is author of *Jokes Don't Jump from Nowhere: Comic Dialogism in P.G. Wodehouse's* 'Jeeves and Wooster' *Novels* (Wissenschaftlicher Verlag Trier, 2012) and editor of *Semiotic Encounters: Text, Image and Trans-Nation* (Rodopi, 2009). Her current work *The Stories We (Like to) Wear* looks at prose narrative and film that raise awareness for the textile industry's exploitation of people and the planet and inspire prosocial and pro-environmental action. After having taught English and American Literature and Culture at the University of Stuttgart since 2007, she is now training to become a secondary school teacher. She is also a voluntary member of the Clean Clothes Campaign, a European network of organisations that work together for justice in the global textile industry.

Joseph P. Schwieterman, Ph.D., a Professor of Public Service Management at DePaul University, is a nationally known authority on transportation and urban development. The director of DePaul's Chaddick Institute for Metropolitan Development, Professor Schwieterman is actively involved in research about planning and development in metropolitan Chicago. Schwieterman's books include *The Politics of Place: The History of Zoning in Chicago*, which was named the best book of regional interest released in 2006 by the Midwest Independent Book Publishers Association. He is one of the founders of DePaul's Sustainable Urban Development graduate degree program.

Enakshi Sengupta, Ph.D., serves as Associate Director of Higher Education Teaching and Learning Association (HETL) and is responsible for the advancement of HETL in Asia, Middle East, and Africa. The Associate Director works closely with the executive director to fulfill the mission of HETL. Dr Sengupta is also the Director of the Center for Advanced Research in Education, Associate Series Editor of the book series, Innovations in Higher Education Teaching and Learning, Emerald Group Publishing. She is the Managing Editor of the *Journal of Applied Research in Higher Education*, Emerald Publishing, and serves as the Vice Chair of the Editorial Advisory Board of the Innovations in Higher Education Teaching and Learning book series, Emerald Publishing. Dr Sengupta is Senior Manager of the Research, Methodology, and Statistics in the Social Sciences forums on LinkedIn and Facebook responsible for managing all aspects of those forums. Dr Sengupta is a Ph.D. holder from the University of Nottingham in Research in Higher Education, prior to which she completed her M.B.A. with merit from the University of Nottingham and Master's degree in English Literature from the Calcutta University, India. Dr Sengupta has previously held leadership positions in higher education institutions.

Yona Sipos works at the intersection of food systems, sustainability, community-engaged scholarship, and equity. She holds a Ph.D. in Integrated Studies in Land and Food Systems from the University of British Columbia, Canada, where she investigated food systems education and long-term partnerships critical for community food systems. Yona now works at the University of Washington as a core faculty of the Nutritional Sciences Program, School of Public Health, helping to launch the new major in Food Systems, Nutrition, and Health.

Martin Stienstra is a Senior Lecturer and Researcher at the Chair of the Netherlands Institute for Knowledge Intensive Entrepreneurship. The chair is embedded in the Business, Management, and Social Sciences Faculty of the University of Twente. Martin holds a Ph.D. in International Entrepreneurship (focus on Effectuation) and an M.Sc. in Business Administration (specialisation International Management), both from the University of Twente. Research topics of interests are, next to effectuation, cognitive decision styles and the influence of national culture on decision-making processes in new venture creation. Martin is one of the coordinators of the minor Crossing Borders and also coordinated its predecessors: International Management and International Exploration. Over the years, he has supervised approximately 100 internships. Furthermore, he is involved in bachelor programs such as Advanced Technology, Business Administration and Creative Technology. He also teaches in and supervises study tour programs. He is the Coordinator for the International Management Track of the Master of Business Administration and has supervised around 200 Bachelor and Master graduation projects. He is also a member of the Program Committee of Advanced Technology.

Manuel Vallée is a Senior Lecturer in the Sociology Department at the University of Auckland. He teaches medical sociology, environmental sociology, sociology of science, and is part of the teaching team for an interdisciplinary sustainability course. His research projects focus on: (1) identifying the social processes through which pesticide use is normalized; (2) elucidating how mainstream sources of medical information obscure the relationship between pollution and disease; and (3) building communities that are just, resilient, and sustainable. To these topics, he brings a perspective that combines a comparative approach with a careful attention to history, culture, and political economy. In 2017, Manuel co-edited the volume *Resilience, Environmental Justice and the City*, published by Routledge Press.

Taisir Subhi Yamin is a Professor of Gifted Education. He has a B.Sc. in Physics, an M.A. in Special Education, and a Ph.D. in Gifted Education and Computer Assisted Learning from Lancaster University in England. He is the recipient of academic prizes and fellowships from Jordan, England, and the USA including a Fulbright Award (1996), Leadership in Gifted Education from the World Council for Gifted and Talented Children (WCGTC) (2013), and the British Council scholarship. Professor Yamin is active in the field of science popularization, and has written 20 books, a large number of articles, chapters in edited books, and research papers. In addition, he has developed a number of training packages to develop productive thinking skills (Creative thinking, critical thinking, creative problem solving, and future problem solving). He is an active member of an impressive list of academic institutions including: the British Educational Research Association, the European Council for High Ability, and Bahrain Association for the Gifted and Talented; Former President, Jordanian Association of Physicists, the National Committee for Gifted Education; and Founder, the Qatari Centre for the Gifted and Talented. He is also the National

Chancellor and Vice President of the International Association of Educators for World Peace. He was a delegate of the WCGTC for about 10 years. In 2002, he was elected as a member of the executive committee of the WCGTC. Professor Yamin was involved in developing UAE project for the gifted and talented in Ulm University. In addition, he has established, in cooperation with Todd Lubart and Sandra Linke, the International Centre for Innovation in Education (ICIE). The ICIE has seven branches in: France, Germany, Sweden, Turkey, Canada, Kenya, and Jordan. In 2005, he was the Editor-in-Chief of *Gifted and Talented International*. In March 2009, he was elected as the President of the WCGTC for four years. More recently, he is the General Director of ICIE in Germany, and Research Fellow at Université Paris Descartes. In the year 2013, he founded, in cooperation with Ken McCluskey the *International Journal for Creativity and Talent Development* (IJTDC). This ICIE publishes the IJTDC in partnership with a number of American and European universities. In 2018, he joined the Faculty of Education-University of Winnipeg, Canada as a Visiting Professor.

Samantha Zi Lin Yeo is a Bachelor of Arts Honors Psychology graduate from the University of Auckland. In 2014, she graduated with a Diploma in Business Studies with a specialization in Marketing at Ngee Ann Polytechnic, Singapore. In 2017, she received a Bachelor's degree in Psychology with a minor in Employment Relations and Organisation Studies from the University of Auckland. Her honors dissertation examined the effects of materialism, environmental concern, and knowledge on pro-environmental clothes buying behavior. It was part of a larger clothing study, supervised by Niki Harré. She is interested in combining her knowledge of business marketing and psychology to help achieve better environmental and human well-being.

Ruthie Yow joined the Center for Serve-Learn-Sustain in 2017. In her capacity as Service Learning and Partnerships Specialist, her work focuses on deepening the capacity of faculty and students to understand and act on equity as central to the creation of sustainable communities. Before coming to Serve-Learn-Sustain, her research and teaching engaged equity at the intersection of race, power, and education in the US Ruthie holds a courtesy appointment in Georgia Tech's School of History and Sociology and has a Ph.D. in American Studies and African American Studies from Yale University.

Yulun (Darren) Zhang has a Bachelor of Science from the University of Auckland. His interest in Green Chemistry encouraged him to take part in the clothing research project under the supervision of Dr Niki Harré. He is interested in the psychological reasons behind clothes purchasing habits and the environmental damages caused by fast fashion. He believes that including sustainability in university education would be a good starting point to encourage a "greener" way of living.

NAME INDEX

SUBJECT INDEX

Printed and bound by CPI Group (UK) Ltd, Croydon, CR0 4YY

13/09/2026

14961059-0002